'The greatest value of the book lies in the specific empirical studies it contains. These can be used in the classroom and also to illustrate general arguments about the relationship between scientific studies of environment and their social context.'

—**Arjun Agrawal**, University of Michigan, USA, American Journal of Sociology

'. . . the absence of abstruse terminology and the fact that the interpretations remain close to the data make these case studies exceptionally well-suited for use as supplementary reading assignments for undergraduate or graduate courses in environmental sociology.'

—**Bill Markham**, University of North Carolina-Greensboro, USA

Also by Steven Yearley

MAKING SENSE OF SCIENCE: Understanding the Social Study of Science

SOCIOLOGY, ENVIRONMENTALISM, GLOBALIZATION

THE GREEN CASE: A Sociology of Environmental Arguments, Issues & Politics

PROTECTING THE PERIPHERY: Environmental Policy in Peripheral Regions of the European Union (*edited with Susan Baker and Kay Milton*)

THE NEW REPRODUCTIVE TECHNOLOGIES (*edited with Maureen McNeil and Ian Varcoe*)

DECIPHERING SCIENCE AND TECHNOLOGY: The Social Relations of Expertise (*edited with Ian Varcoe and Maureen McNeil*)

SCIENCE, TECHNOLOGY AND SOCIAL CHANGE

SCIENCE AND SOCIOLOGICAL PRACTICE

Cultures of Environmentalism

Empirical Studies in Environmental Sociology

Steven Yearley
Edinburgh University

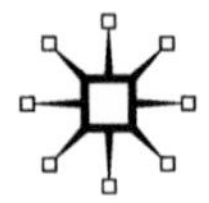

First published in hardback 2005
This paperback edition published 2009 by
PALGRAVE MACMILLAN

Palgrave Macmillan in the UK is an imprint of Macmillan Publishers Limited, registered in England, company number 785998, of Houndmills, Basingstoke, Hampshire RG21 6XS.

Palgrave Macmillan in the US is a division of St Martin's Press LLC,
175 Fifth Avenue, New York, NY 10010.

Palgrave Macmillan is the global academic imprint of the above companies and has companies and representatives throughout the world.

Palgrave® and Macmillan® are registered trademarks in the United States, the United Kingdom, Europe and other countries.

ISBN-13: 978–1–4039–0120–0 hardback
ISBN-13: 978–0–230–23711–7 paperback

This book is printed on paper suitable for recycling and made from fully managed and sustained forest sources. Logging, pulping and manufacturing processes are expected to conform to the environmental regulations of the country of origin.

A catalogue record for this book is available from the British Library.

Library of Congress Cataloging-in-Publication Data
Yearley, Steven.
Cultures of environmentalism : empirical studies in environmental sociology / Steven Yearley.
p. cm.
Includes bibliographical references and index.
ISBN 978–1–4039–0120–0 (cloth) 978–0–230–23711–7 (pbk)
1. Environmentalism—Social aspects. 2. Environmental policy. I. Title.
GE195.Y43 2005
333.72—dc22 2004051672

10 9 8 7 6 5 4 3 2 1
18 17 16 15 14 13 12 11 10 09

Transferred to Digital Printing in 2009

Contents

List of Figure and Tables

Figure

Tables

1
Introduction: Studying Environmental Issues Sociologically

The day after I send the manuscript of this book to the publishers, some friends have promised me a boat trip from Queenscliff, a Victorian seaside resort. As well as the boat ride there will be a chance to visit the Ozone Hotel, a venerable establishment dating from the late nineteenth century. The hotel took its name from a paddle steamer that used to run day jaunts to Queenscliff, outings that were believed to offer urban trippers the chance to breathe in some invigorating – ozone-rich – sea air. On my day trip, I will be concerned with ozone in a different way. Because of the recent depletion of the ozone layer, I will have to take care to apply extra sun-cream to offset the effects of increased ultra-violet radiation. In about a century, ozone has gone from a blessing into an anxiety and somehow the Ozone Hotel's name has retained its topicality. Just as in the naming of the paddle steamer and the hotel, we inscribe our views and concerns about the environment onto the cultural objects around us. The environment suffuses society and our treatment of cultural objects reflects back on our understanding of the environment itself. This basic observation lies at the heart of sociological approaches to the environment. Environmental sociology asks where our views of nature and the environment come from, how our conduct towards the environment is moulded, how we draw the distinction between nature and culture, and how our knowledge of the environment is shaped. The studies in this book are designed as empirical investigations of these themes, of various cultures of environmentalism.

Sociology is about as old as the Ozone Hotel, but environmental sociology is only a recent invention. In the 1980s, when I began my work on environmental organisations there was a good deal of environmental activism but hardly any environmental sociology. Now the reader has a wide choice of books. Compared to most of its rivals, this book is meant to stand out because it is primarily empirical. It is a collection of studies of environmental campaigns, disputes, movements and problems. As sociological interest in the environment has grown, sociologists have sought to bring environmental issues within the scope of their theories. This is a welcome development. However, it has meant that it is now possible to write books about the environment for a sociological audience that are primarily organised around theoretical schools and the debates between them. I have tried to do something different by keeping the theory in the background and highlighting the details of the cases. I hope this strategy makes the chapters accessible and appealing. But I believe it serves a further function by focusing attention on the question of how it is that I claim to know the things that the studies describe. Though – in the interests of readability – the chapters do not have lengthy methods sections I have tried to foreground issues of evidence by, for example, including lengthy quotes from interviews and describing the settings in detail. In this sense, the chapters are simultaneously about the environment *and* about one way to undertake sociological studies of the environment.

The book is divided into three parts, each composed of three chapters. Part I, 'Cultures of Movement: The Sociology of Environmental Movements and Problems', examines the culture of environmental campaigning and activism. In particular it looks at how the environmental movement differs from other contemporary social movements and asks how the culture of environmental organisations influences the nature of environmentalism itself. The issue of the shaping of environmentalism is then developed through studies of the ways in which environmental campaigning and concern have been refashioned in response to the globalisation of environmental issues. Part II, 'Studies of Environment, Law and Public Policy', consists of three UK case studies examining how policy and planning decisions are made. These studies demonstrate how legal and governmental institutions – 'modernist institutions', theorists would call them – attempt to handle environmental matters. My aim is to show that detailed analysis of environmental case studies allows sociology to focus on pivotal issues in the way that contemporary cultural institutions try to fit modernist conceptual tools to the evaluation of culture and nature. In Part III, 'Cultures of Knowing and Proving: Science, Evidence and the Environment', I investigate the role of knowledge about the environment. Sunderlin (2003) and other analysts have rightly noted that environmental issues are deeply ideological in the sense that where some commentators are struck by evidence of imminent environmental catastrophe others see nothing but continued progress, dependent only on sensible and affordable adaptations to ecological change. So entrenched have these positions become that, at the start of the twenty-first century, attempts to make an overall assessment of the state of the environment have come to be as much a right-wing as an environmentalist or left-wing preoccupation. None the less, environmental campaigns and policy decisions frequently turn on particular claims to knowledge – that genetically modified (GM) crops can safely be farmed (or not) or that climate change is (or is not) being caused by industrial emissions. Accordingly, in Part III, I employ recent case-study material to examine in close detail how knowledge about the environment and environmental threats is developed, contested and determined.

The book concludes with a separate chapter that belongs to none of the parts: 'The Value of Environmental Sociology: Towards a Sociology of the Sustainable Society'. In Chapter 11, my aim is to use the conclusions from the nine empirical chapters to examine the contribution sociology can make to our thinking about environmental sustainability. In short, the chapter claims that, to date, ideas about sustainable living have been limited because they have neglected the sociological aspects of the sustainable life. In the chapter, I identify three leading sociological dimensions of sustainability. In this sense I believe that environmental sociology is not only a way of understanding contemporary social change but a key part of the enterprise of trying to see beyond present practices and techniques to the sociology of the future.

Of the following ten chapters, five are newly written and five are updated versions of previously published studies. I am very grateful to the publishers for their permission to present this work again in a new context. First, I would like to thank Sage Publications Ltd for permission to reprint the articles that make up Chapters 5–7 (originally Yearley 1989, 1992b and 1999a): 'Bog-standards: science and conservation at a public inquiry' in *Social Studies of Science*, Vol. 19, pp. 421–438, Sage Publications, 1989; 'Skills, deals and impartiality: the sale of environmental consultancy skills and public perceptions of scientific neutrality' in *Social Studies of Science*, Vol. 22, pp. 435–453, Sage Publications, 1992; and 'Computer models and the public's understanding of science: a case study analysis' in *Social Studies of Science*, Vol. 29, pp. 845–866, Sage Publications, 1999. Secondly, I would like to record my thanks to the publishers of the *British Journal of Sociology* for permission to reprint Chapter 8 (originally Yearley 1992c). Thirdly, my thanks go to the Athlone/Continuum Press for permission to use the publication by

Steven Yearley and John Forrester, 'Shell, a sure target for global environmental campaigning?' in Robin Cohen and Shirin Rai (eds), *Global Social Movements*, London: Athlone 2000, pp. 134–145, as Chapter 3 (originally Yearley and Forrester 2000). The original version of Chapter 3 was jointly authored with my colleague John Forrester and I am particularly grateful to him for allowing our work to be published in this book; he is the co-author of Chapter 3. All errors and oversights are, regrettably, down to me; not even John can be asked to share the blame.

Part I

Cultures of Movement: The Sociology of Environmental Movements and Problems

2

Social Movement Theory and the Character of Environmental Social Movements

Introduction: environmental sociology and the study of social movements[1]

The end of the millennium provided an apparently irresistible occasion for trying to take stock of all kinds of human endeavours. Commentators asked whether war and violence were decreasing or on the rise; the Jubilee 2000 pressure group proposed that this was a fitting time for the forgiveness of international debt; religious spokespersons wondered about the spiritual state of humanity at this symbolic date, symbolic in the Christian world at least. But it was also a critical date for assessments of environmental progress. During the 1990s many environmental groups, convinced of the rapidity of ecological deterioration, had set the year 2000 as some kind of milestone. But in the end, no enormous environmental changes occurred to mark that date. In the market, all was mostly business as usual; the stock exchanges were all riding high. Nor had the unequivocal environmental depredations they had feared become overwhelmingly apparent. Yet, at another level there was a change since, by the end of the millennium, social movement organisations (SMOs) had become the most popularly acclaimed and, in many respects, trusted agencies advocating large-scale environmental change. They had won widespread public admiration because of their daring and heroic undertakings, because of the verve and symbolic acuity of their actions and because they seemed to be in the vanguard of environmental change and to be responding to the challenges of globalisation. Of course, commentators noted that governments and inter-governmental agencies might have more power to set and influence environmental standards, that companies might be making the greatest impacts on the environment, that it was often scientists who identified possible environmental problems which were 'off the radar' of environmental groups, and that the daily consumer choices of the industrialised world's massed citizens and commuters might outweigh their efforts. All the same, social movements represented the quintessential environmental actor. In cultural terms, environmental organisations stood for the environment in a way which Environment Ministers, the collected scientists of the Intergovernmental Panel on Climate Change (IPCC) (see Chapter 4) or corporate donors such as Shell simply could not.

At the same time, social movements increasingly commanded the attention of social scientists and commentators. First, social movements and the associated movement organisations appeared to confound expectations. Far from politics as usual, social movements indicated how successfully and how enduringly people could be organised – or organise themselves – around non-conventional political objectives. Standard economic and political theories did not anticipate that people 'ought' to mobilise so successfully around a diffuse political objective such as global environmental improvement. In the UK, where movement organisations succeeded even more thoroughly than elsewhere in institutionalising themselves, several groups had memberships in the hundreds of thousands, with the largest easily exceeding memberships of political parties and coming well within an order of magnitude of the total membership of the Union movement.

Secondly, social movements appeared effortlessly and beguilingly innovative: they pioneered new forms of campaigning designed to maximise publicity. Greenpeace have perhaps become the masters of this practice, generating widely circulated images of high-seas heroics. In the case of the Brent Spar protests in 1995 (aimed at preventing the ocean disposal of an oil-storage platform, see Chapter 9), so successful were they in their command over the production of news images that the terrestrial television industry was prompted to reflect on its information serfdom. Virtually all the video coverage of the controversy surrounding the ageing oil installation was supplied by Greenpeace itself. Comparable creativity was shown by UK anti-roads-protesters in the mid-1990s who developed techniques for delaying highway construction first by occupying the trees which were to be felled to make way for the road, then by tunnelling under the path of construction equipment. Tree camps powerfully symbolised the solidarity between protesters and the aspects of the natural world they were trying to protect, but the tunnels represented a special daring since it was the very fragility of structures which ensured that constructors (probably) would not dare drive construction equipment over the occupants. Jasper has recently written of the *Art of Moral Protest* (1997) and it is this kind of phenomenon whose artfulness he salutes. In recent years, other environmental activists have shown similar creativity, whether reclaiming the streets by holding fairs on motorways or cycling in concert to disrupt the flow of motorised traffic.

Finally, social movements appeared effective; many campaigns with which they were associated (even if they did not initiate them) were successful, including limitations on whaling, acid rain reduction, regulation of ozone-depleting chemicals in Europe, the protection of numerous endangered species from trade or hunting, and limitations on sea-dumping. Yet, despite the growing social scientific and journalistic interest in social movements, there is no agreed way of analysing the nature or the functioning of these environmental SMOs. On the contrary, competing and conflicting interpretations persist. In this chapter, I aim to consider how the understanding of these SMOs can be extended by employing recent work in the sociology of social movements to analyse the characteristic features of the environmental movement. At the same time, I propose to evaluate the competing analytical approaches in terms of their ability to throw light on the workings of the environmental movement. The chapter will consider which features of environmental SMOs are common to other social movements and which are not; the identification of the movement's sociological peculiarities will be of particular interest. Given that the environmental movement and movement organisations have been highly influential in bringing about changes in public attitude and commercial behaviour, in influencing public policy and in motivating people to engage in practical projects in the environment's perceived interest (for a breathless catalogue, see Fred Pearce 1991, ix–xi), there is also a practical interest in getting the diagnosis correct. Sociology needs to refine its apparatus for describing these groups and this movement.

In the early 1990s, Newby attacked sociology and, in particular, British sociology for making only a 'slender' contribution to the study of the environment (1991). Judging from Newby's comments one might suppose that the researcher would be hard-pressed to find sociologists writing at that time about environmental groups and ecological issues. On the contrary, many authors had already sought to include green groups within the compass of their analysis. And those authors represented a wide range of sociological perspectives. At one extreme, leading theorists such as Giddens had tried to incorporate the green movement within their large-scale interpretations. Giddens, exploring what he terms high modernity, emphasised the ubiquity of risk and danger in the modern world. Among the global risks that shape contemporary culture and politics, he gives prominent position to 'ecological calamity and uncontainable population explosion' (1990, 125). For Giddens, green movements constitute one of the four types of social movements, which represent a response to risks in the principal 'institutional dimensions' of modernity. He sees 'clear countertrends, partly expressed through ecological movements', challenging the logic of relentless, institutionalised technological innovation (1990, 170). In his work on the 'risk society', Beck went even further by nominating ecological protests as a prime example of what he terms reflexive modernisation, the destructive application of modernist principles to themselves (1992, 156). By using scientific methods to criticise the harm caused by our scientific civilisation – in particular by using scientific reasoning to calibrate the uncertainty and risk associated with agrochemicals, nuclear power and additives – environmental campaigners publicly expose the self-reflexive fix into which 'late modernity' has got itself.

At a rather less abstract level Melucci, who regards social movements as in many respects characteristic feature of complex modern societies, also placed considerable emphasis on green movements. For him, 'the peace and ecological movements [are key] precisely because they are testaments to the fragile and potentially self-destructive connections between humanity and the wider universe' (1989, 6). Additionally, environmental groups figured in many case studies of the formation and development of social movements. This is true both within the US tradition, which concentrates on the formal and organisational characteristics of social movements (see Zald and McCarthy 1987, 179 on the Sierra Club and Friends of the Earth (FoE)), and within the European mould, which seeks to relate social movements to political and class interests (for example, see Touraine 1983 on the French anti-nuclear movement). These studies have also been complemented more recently by exploratory studies of new forms of social organisation and protest, for example anti-roads protests in Britain (McKay 1996), and by attempts to interpret the rapidly developing environmental justice campaigns, primarily in the USA (see Bullard 1990; Ringquist 2000). Lastly, in work which largely appeared later in the 1990s, attempts have been made to conceptualise the nature of international collaborative movement activity, using terms such as 'transnational advocacy networks' (Keck and Sikkink 1998).

Accordingly, my task in this chapter is in no sense to initiate sociological interest in environmental social movements but to clarify and systematise existing work. I shall attempt this under two main headings: the theoretical interpretation of environmental social movements and the characteristic features of the environmental movement. In the latter section, I shall draw on comparisons between environmental and other progressive movements to identify organisational features that are especially marked in the environmental movement's case.

Interpreting environmental social movements

Virtually every commentator on the sociology of social movements makes use of the distinction, which I have already mentioned, between the US and the European interpretative traditions. On the European side are those who assess social movements primarily in relation to their perceived capacity for major social transformation. While there may be many campaigning groups or sets of people agitating for social change or alterations in public policy, only some of these qualify for the appellation 'new social movements' (NSMs). Such movements are described as 'new' essentially because they cannot be seen as directly deriving from the 'old' labourist movements. In other words, early social analysts had assumed that social movements were essentially expressions of the accepted, 'major' political forces; they were seen as embodiments of workers' attempts to further their economic and political interests and therefore derivative in some way from class politics. However, the NSMs did not seem to reflect this dynamic. The student movement, civil rights and peace movements and even the women's movement seemed to defy earlier assumptions, drawing support across the classes and pursuing political ends which could not be boiled down to obvious class objectives. Often, participation in, or support for, such movements was more closely correlated with educational experience than with socio-economic class position.

Although not unified by presumed class alignment, the commonly listed NSMs share a progressive political orientation. For example, the women's movement, civil rights groups and radical environmentalists can all be said (approximately at least) to be extending our conception of rights: to women, to formerly excluded minorities and to the natural world. For this reason, the NSMs are seen as the natural allies of the 'progressive' left. And some analysts, despairing that the working classes were ever going to bring about a social transformation, transferred their hopes for radical reform onto these movements. In this way of thinking, the social movements can take on the historic role usually ascribed to classes. As Touraine expresses this idea, 'Social movements are neither accidents nor factors of change: they are the collective action of actors at the highest level – the class actors – fighting for the social control of historicity, i.e. [of society]' (1981, 26).

According to such a view, the only movements that qualify for the title 'new social movements' are ones which take over the historic agenda of social classes and which strive to control the 'great cultural orientations' of society. The problem then is deciding which movements qualify. Civil rights movements might well, but how about the youth movement? What separates a 'great cultural orientation' from a less-than-great one?

The leading alternative, North American, approach adopts a much more disinterested interpretation of a social movement. For this approach, a social movement is defined in terms of its organisational characteristics. Social movements are instances of collective behaviour which are more organised than protesting crowds or mobs, less formalised than political parties, and more concerted than simple social trends. Obviously, the boundaries around this phenomenon are not clear-cut but a reasonable amount of agreement can be reached about what is a movement, what is a party and what is a crowd.

Adherents of this view note a long-term trend. Early on, social movements tended to be made up of massed participants. More recently the role of professionalised cadres of movement organisers has become markedly significant. As early as the 1980s, the apparently most vociferous social movements were actually highly efficient SMOs, sometimes followed by few spontaneously active supporters. Analysts drew attention to the importance of organisational and technical innovations such as the pre-printed letter addressed to politicians or senior officials and, especially popular in the USA, records of the voting performance of state and federal politicians. A politician who had once voted for, or made favourable remarks about, say abortion or gun control was likely to have his or her record resurrected by right-wing pressure groups. Such methods were soon supplanted by electronic techniques, by means of which many thousands of messages can be sent to leading politicians or to government departments in a co-ordinated way with very little effort from the 'participating' public. Employing these types of methods, professional campaigners can readily create the impression of widespread social concern and thus of a large social movement even while ordinary supporters of the proposed change have to do remarkably little.

Such techniques, which represent a rationalisation of social movement practices, became widespread, adopted by moral crusaders, wildlife conservationists and those opposed to European federalism. Analysts within this North American tradition are accordingly struck more by the similarity of groups irrespective of their ideological affiliation than by differences between those directed at 'cultural transformation' and those dealing with more sectional, sectarian or even trivial concerns.

However, the US-style approach, which emphasises the 'moral entrepreneurship' of figures within SMOs, suffers one clear drawback. It seems unable to distinguish between effective lobbyists or pressure groups and broader social movements, where the latter offer – at least in principle – the potential for mass participation in some form. The peace movement would be able to take advantage of the most sophisticated forms of direct mailing and email-messaging and to welcome payment by credit card or direct bank debits whilst practising canny lobbying of industry and government; at the same time its mass membership is central to its character. It is difficult to conceive a similar situation developing for a group campaigning on the arts – say, in support of opera provision. The social attributes of many supporters might be similar, the methods adopted by the organisations' professional staffs closely related, but one cause has been associated with a participatory movement and the other has not. Approaches that focus almost exclusively on organisational characteristics tend not to illuminate the distinction between 'promotional' pressure groups (groups founded to advance a principle or cause) and those SMOs associated with a broad movement. Environmentalism certainly qualifies as a broad movement; even in Britain the movement is estimated to have had up to four million supporters during most of the 1990s (McCormick 1991, 152). Accordingly, one needs to ensure that this distinction is explicitly analysed and accounted for.

In summary, the nominalist, US approach tends to be too undiscriminating while the European analysts' view is inclined to be excessively restrictive. But this overview of the two competing traditions at least allows us to identify key descriptive elements that characterise a social movement: it has a large-scale membership and a promotional character; modern social movements will tend to be associated with sophisticated administrative apparatuses.

The question then becomes, how can the analyst identify and account for this kind of social phenomenon? It seems to me that there have been four principal sociological responses. First, there is the 'Continental European' response which tends to be Hegelian. That is, 'true' social movements are viewed in a historicist way, with the assumption that they articulate the interest of some developing historical actor. As I mentioned above, this impression is easy to reinforce since many social movements share a 'progressive' orientation. But there appears to be no non-arbitrary way of separating the 'true' from other movements which possess all the characteristics of the 'true' ones apart from a progressive political basis. For example, aside from their profoundly anti-universalistic outlook, nationalist movements would seem to fit the bill perfectly well. They have frequently been transformative and seem to have enjoyed large-scale historical consequences. Nothing, apart from the analyst's prior assumption about the political orientation of 'true' movements, could disqualify them from movement status.

A second approach to a systematic classification is illustrated by Giddens' identification of the leading 'institutional dimensions of modernity'. Giddens argues that there are grounds for identifying four relatively independent dimensions along which modernity develops; for Giddens (1990, 59) these are capital accumulation, surveillance, military power and industrialism (the last of these representing the 'transformation of nature'). Giddens' suggestion appears to be that, having identified these axes, one can determine whether social movements have arisen in response to each. A plausible-enough case can be made for saying that they have: respectively, the labour movement, democracy/free speech movements, the peace movement and the ecology movement. If it could be established that these are indeed the principal axes of advanced societies, then there would be some intuitive support for the idea that the social movements which crystallised around them are of special relevance; they would be the 'historic' ones. However, as Giddens himself notes, the 'objectives of feminist movements... crosscut the institutional dimensions of modernity' (1990, 162). In other words, one of the leading progressive social movements, and arguably the one with the most influence in sociological circles, fails to conform to his theory. Thus there is no neat fit between the dimensions of modernity and the dimensions of SMO. The same kinds of difficulty confront analogous Habermasian analyses which attempt to categorise types of social movement. In short, the specification of the essential characteristics of modern capitalism has not been consensually achieved so we cannot 'check off' the movements against the corresponding features of capitalist society. Nor, in any case, does the present variety of social movements map readily onto the leading theorists' anatomies of present-day capitalism.

A third theoretically based attempt to put forward systematic grounds for separating social movements from 'mere' pressure groups and other instances of collective behaviour has been made by Eyerman and Jamison. Their argument was that it is central to see that social movements are engaged in what they term 'cognitive praxis'; in other words, social movements are characteristically producers of innovative knowledge claims. They suggest that 'A social movement is not one organization or one particular special interest group. It is more like a cognitive territory, a new conceptual space' (1991, 55). Furthermore, 'our approach tends to limit the number of social movements to those especially "significant" movements which redefine history, which carry the historical "projects" that have normally been attributed to social classes' (1991, 56).

These authors, who were also centrally involved in a three-country empirical study of the environmental movement (Jamison *et al.* 1990), are quite correct to stress the fact that social movements commonly advance new knowledge claims and that movement intellectuals can play a decisive role. As I shall argue when I come on to the part played by scientific authority in the green movement, the fact that environmentalists' arguments depend on science has had strong sociological implications for the profile of the movement and the make-up of its member organisations. But it appears premature to make cognitive praxis a defining characteristic of 'true' social movements. This is both because it is easy to see politically significant social movements featuring little cognitive innovation (the moral majority, nationalism) and because there are several abortive instances of cognitive praxis (assorted utopias, even eugenics). The latter groups defined novel 'conceptual spaces' but failed to give rise to enduring movements. Cognitive praxis alone does not a movement make, a fact which is demonstrated by the way in which Eyerman and Jamison end up fleshing out their theory in an *ad hoc* fashion, asserting that 'only those [social problems] that strike a fundamental chord, that touch basic tensions in a society have the potential for generating a social movement' (1991, 56). In the end, even for Eyerman and Jamison, cognitive praxis is not the final explanation for social movement formation; rather it is a matter of basic tensions. However, they do not elaborate on how one can identify or analyse 'basic tensions' or 'fundamental chords'; this explanatory resource remains uninvestigated. There is, no doubt, a strong association between social movements and cognitive innovation. But it seems unduly restrictive to make such innovation a defining characteristic.

This, I suggest, leaves us only with an empirical generalisation, an essentially descriptive definition of social movements. Social movements are to be known by their organisational forms; although they are often progressive, they need not be; although they often engage in cognitive praxis, they need not do so. There is, in other words, no single criterion allowing us to explain which pressure groups can be associated with a large-scale social movement and which cannot; the reasons for mass participation have to be analysed on a case-by-case basis. It is easy to sympathise with sociologists' pursuit of the one key to identifying the characteristic features of SMOs. Simplicity in theoretical analysis is widely valued across the sciences. But staking all on a simple scheme when it does not wholly work is not a shrewd move. Accordingly, though it lacks the theoretical clarity which the competing interpretations sought, the approach advocated here has the benefit that – since it lacks one single key element – it draws attention away from the supposed unitary character of movements. Thus, if we examine the array of environmental SMOs in Britain, we can see that it is difficult to speak of *a* movement – in fact, the term 'social movement industry' employed by US analysts is highly apt. Although the leading SMOs collaborate over specific campaigns and share supporters, and although there are frequent meetings between the different professional cadres, there is competition between them: competition for the highest profile campaign topics; competition for acceptable and wealthy backers; competition for news coverage; and – particularly in recessionary times – competition for members and their money. Given this atmosphere of competition, Zald and McCarthy (1987, 179) record their surprise that little attention has been paid to 'inter-SMO' relations. Drawing parallels with the behaviour of firms and formal organisations, they formulated 14 generalisations about SMOs' behaviour in competitive situations. In the light of such competition within a movement and of the fact that such competition is largely played out between SMOs (professionalised organisations keen to retain staff numbers, with staff members reasonably keen to hang on to their own jobs), any sociological approach that insists on analysis only at the level of whole movement will inevitably appear superficial. Phenomena below the level of the whole movement will be highly consequential. Thus, the outcome of the competition between SMOs will shape the future direction of the movement; such micro-level features will, for example, influence the future 'cognitive praxis' that comes to predominate within the movement.

Furthermore, approaches that are set on identifying the key features of 'true' social movements run the risk of diverting attention away from changes in the political and legal environment in which SMOs operate, changes that may alter the fundamental way these organisations conduct themselves. For example, as Greenpeace scored campaign successes through the 1980s and 1990s it became wealthier and able to afford more equipment, employees, boats and property. This made it susceptible in a way that it would not previously have been to novel legal manoeuvres introduced by the UK authorities that allowed the sequestration of the assets of groups – originally trade unions – which did not comply with legal injunctions. Now the targets of Greenpeace campaigns had a better prospect of suing the organisation in relation to the losses they were suffering and Greenpeace had many more identifiable 'goods' against which claims could be made. With more to lose, and new powers available to the authorities, Greenpeace became more circumspect, apparently changing its 'essential' characteristics as an SMO. This change was driven not by the character of the movement itself but by changes in the political context.

A second reason for being wary about describing the environmental lobby in Britain in anything like the Tourainean sense is that the groups which are both radical and effective tend to be run as centralised organisations. Greenpeace is famously so, leading Allen, a campaigning journalist working on waste incineration, to complain that, in Britain, 'Greenpeace is very definitely bureaucratic and was seduced by the establishment fairly quickly. From a small grouping at the beginning of the eighties, Greenpeace displays all the trappings of a multinational company or a civil service department' (1992, 223). How rapid a seduction has to be to qualify as 'fairly quick' is unclear to me, and – in any case – Greenpeace continues to engage in direct actions. But Allen's apparent resentment stems from the alienation sometimes felt by community campaigners in the face of professionalised and wealthy campaign organisations (on the US experience, see Dowie 1995). McCormick notes that into the 1990s Greenpeace became 'less confrontational, and more inclined to use the same tactics of lobbying and discreet political influence once reserved by the more conservative groups' (1991, 158). However, as the case studies in Chapters 9 and 10 indicate, this judgement was rather premature as Greenpeace sought to recapture the moral high-ground afforded by carefully chosen direct-action campaigns (Bennie 1998). None the less, the key point is that movement organisations may have an uneasy relationship with participants in the movement. The SMOs and the movement are not the same thing and may even be at odds (see della Porta and Diani 1999).

In Britain, FoE attempted to institutionalise an arrangement for cooperation between a centralised London-based staff and its local groups. The latter can select their own campaign targets but are bound by a licence agreement with FoE, which prevents them acting in FoE's own name. Local groups have a form of parliament at the annual meeting but cannot require the board to change its policy. The headquarters staff has a highly professionalised ethos (McCormick 1991, 117–18). The structure lessens but does not remove tensions between the core and the regionally active members. Even FoE, which has tried to build in a mechanism for sustained member participation, cannot be viewed as the expression of *the* movement.

Finally in relation to the nominalist, descriptive approach, it should be noted that its exponents are not denied access to explanation at the level of macroscopic social change. Thus, Berger has argued for the relevance to progressive SMOs of the rise of the 'knowledge class', a 'new middle class . . . of people whose occupations deal with the production and distribution of symbolic knowledge' (1987, 66). These are the intellectuals and service workers whose knowledge is not generally directed towards material production; they work in education, counselling and communications and in the 'bureaucratic agencies planning for the putative nonmaterial needs of the society (from racial amity to geriatric recreation)' (1987, 67). This knowledge class tends, in Berger's view, to be antagonistic to the core values of capitalism, a fact which he attributes chiefly to two factors: the interest this class has in having 'privilege based on educational credentials'; and the vested interest it has in expanding the role of the welfare state where its members find work. As he expresses it, '[The] knowledge class has an interest in the distributive machinery of government, as against the production system, and this naturally pushes it to the left in the context of Western politics' (1987, 69). Through analyses of this sort one can make sense of the typical constituency of support for environmental movements, movements that tend to appeal to the well-educated and to those in the distributive aspects of the economy. We can, in other words, understand the principal customer base to which environmental SMOs appeal. But as Berger points out, the knowledge class is huge. It forms much of the basis for support for the green movement but knowledge-class membership under-determines the propensity to lend support. Other people with an identical class profile to that of keen environmentalists may support different movement causes, promoting community arts or alternative therapies.

In summary, no doubt Melucci (1989, 194) is partly correct to ascribe the differences between the typically European and the typically US analytical approaches to differences in the social and political context in the two continents. Until the 1990s at least, political activism appeared more pluralistic in North America and more class-based in Continental Europe. But the intuitively appealing proposition of the European tradition that only some historic movements ought to be distinguished by the title of 'new social movements' appears to be unsustainable. No theoretical or methodological basis can be offered for identifying the exclusive claimants to this title. The environmental movement surely merits the title of a new social movement. But, while at first sight it appears to have the correct kind of transformational potential to qualify for movement status in the European sense, its transformational ability remains to be empirically demonstrated. Furthermore, competition and divisions among environmentalists and between SMOs also imply that its coherence as *a* movement must be in doubt.

Comparing environmentalism with other New Social Movements

By arguing that the term 'new social movement' should be taken as an empirical generalisation and that the term should be used to identify a group of organisationally similar social phenomena, one arrives at an analytical position which invites comparisons between various social movements. Since they are not assumed, in the Hegelian manner, to share essential features in common, it remains an open empirical question how they compare with one another. In this section, I shall concentrate on three dimensions on which the environmental movement stands out in significant empirical ways from its close relatives. The first characteristic, which will be discussed at greater length in Part III of this book (Chapters 8–10), concerns the role of scientific evidence, argumentation and expertise within the environmental movement. The role of scientific authority in the green movement is easily illustrated at a practical level. More traditional nature conservation groups, such as the Royal Society for the Protection of Birds (RSPB, the largest conservation or environmental group in Europe), have sizeable scientific budgets. The RSPB engages in research on bird migration, on bird ecology and on the impact of farming practices on wildlife as part of its core activities. More campaigning organisations, such as the World Wide Fund for Nature (WWF) and even Greenpeace, also devote resources to scientific analysis, as demonstrated when Jeremy Leggett was appointed Director of Science at Greenpeace (UK). It is reflected in the growth of a popular science literature among the interested public, with books such as Rachel Carson's *Silent Spring* reaching very large audiences. It is demonstrated by the growing commitment of national governments to spending on environmental research, on centres for the study of climate change, biodiversity conservation and atmospheric monitoring. But the importance of this scientific activity has not, by and large, been remarked by social movement analysts, except at the level of generality at which theorists such as Beck write (see Yearley 1992a, 113–48). For example, it receives only passing and unsystematic attention in Dryzek *et al.*'s recent comparative study (2003, 84–86).

Of course, one might argue that technical expertise has not been emphasised because it is common to other social movements or pressure groups which also make extensive use of scientific arguments. Public health organisations of the nineteenth century would be a clear example; they used new ideas about the spread of disease and the importance of hygiene to try to change people's behaviour and public policy. Candidate NSMs from the twentieth century are not without their scientific aspects either. Thus the idea of a nuclear winter became an important part of peace campaigners' case in the 1980s. It was argued that multiple strikes from nuclear weapons would cause such a build-up of dust and particles in the atmosphere that an enormous extinction would take place. Crops would not grow, animals would die. It would be like the aftermath of those prehistoric meteorite strikes which are believed to have been responsible for mass extinctions in the past (for example, the extinction of the dinosaurs) primarily through changing the climate and obscuring the sun for lengthy periods. Campaigners argued that the idea of a winnable nuclear war was undermined by this scientific demonstration that a war would cause worldwide extinction. Similarly, a robust attitude to biologists' arguments was central to the development of feminism in the 1970s. Where biologists tended to see natural sex differences, feminists saw social and cultural differences erroneously imposed onto nature by conventionally minded biologists (see Oakley 1972).

Despite these points, it appears that science is of even greater importance to the environmentalist case. For one thing, many of the objects of environmental concern are only knowable through science. Without a scientific worldview we would know nothing of the ozone layer and would certainly be unable to measure its diminution; the same is true of the greenhouse effect. But science is important in other ways too. Natural history instructs us in the needs of animals and plants (what or whom they need to eat, where they like to nest, the hazards to which they are susceptible). Environmental management, of nature reserves or of elephant populations, depends on scientific understandings. This is not to say that science has been a dependable friend to the environmental movement. In some ways the contrary is true. As is well known, the demand for scientific 'proof' has been used to justify official inactivity towards environmental problems (over the abatement of acid air pollution for example); the authorities have argued that they should wait until the evidence of damage is conclusive before making major policy changes. The scientific community, particularly in the nuclear field, has often appeared committed to activities that heighten environmental hazards; more recently the same appears to have held true for the biotechnological research community (see Chapter 10). Threats to the environment have often stemmed from technological progress; after all it was scientific advice that prompted the addition of lead to petrol and gave rise to the widespread use of CFCs (chlorofluorocarbons, the family of chemicals that caused most damage to the ozone layer).

Since both sides in environmental debates typically seek to enlist the support of scientific arguments, it makes little sense to try to say which 'side' science is on. However, one can note that the central role of scientific argumentation in the environmental case has had an important impact on the sociology of the movement. First, the importance of scientific expertise has added to the distance between the professional cadres in the SMOs and the casual movement supporter. While campaign professionals learn more and more about acid pollution or radioisotopes, 'ordinary' members are encouraged to deal chiefly in terms of general guidelines such as the 'polluter pays' or 'precautionary' principles. One should not make too much of this observation. Clearly full-time peace campaigners became experts on aspects of nuclear weaponry and on 'Star Wars' technologies, while many 'ordinary' greens are very knowledgeable about ecological issues. But the tension between movement organisations' commitment to the ideal of getting the technically correct answer and the need to listen to their members' views and priorities has led all of the leading environmental movement organisations to struggle with issues of internal democracy. Democracy implies the sovereignty of members' opinions, and this is at odds with campaigners' professional expertise. In a sense, the development, during the 1990s, of highly activist groups such as EarthFirst! – groups that kept their analysis simple, and prioritised action over policy-analysis – can be seen as a reaction to the limitations imposed by movement organisations' expertise.

Second, the centrality of scientific evidence held out the beguiling possibility of winning one's case simply through argument. If one could demonstrate that pollution was eating away at the ozone layer or that emissions from burning fossil fuels were causing climate change, then one might think one's argument would be easily made. It would not be a question of opinion but of demonstrable fact. This notion appealed in particular to the more 'establishment' environmental groups, such as the Royal Society for Nature Conservation (RSNC, now known as The Wildlife Trusts) which had, throughout its 90-year history, had a predominantly scientific ethos. The supposedly universalistic characteristics of scientific argument thus lent credibility to one style of campaigning and have left their mark on the composition of the overall green movement. This factor plays out interestingly in the case of environmental justice movements. Their strategy tends to be to point to a misleading universalism assumed in scientific analyses. For example, they may point out how disadvantaged communities are exposed to multiple pollutants, the combined effects of which have typically not been studied scientifically. Or they may point to differences in lifestyle from one community to another that result in different environmental exposures. For example, people's exposure to environmental toxins may relate to diet and the characteristic dietary practices of various ethnic sub-cultures are typically not reflected in standardised measurements of average exposure. Mainstream environmental SMOs too have been caught out by their inattention to these considerations.

If the environmental movement appears exceptional among NSMs on account of its relationship to science, it also offers to stand out because of the movement's international character. Many social movements have made claims to international solidarity, 'international socialists' being one obvious and somewhat ironic example. In principle there are probably good reasons for international solidarity in all sorts of social movement; for example, peace movements have a strong and urgent reason for international solidarity even if their unity generally foundered on the asymmetrical opportunities for peace organisation during the Cold War. But environmental issues are, I suggest, unusually international for several reasons.

First, environmental threats are themselves often 'transboundary' in character. Whoever pumps them out, CFCs will affect the ozone layer. The USA or Europe cannot attend to its own interests in this matter (although this is not to say there are no disputes over the North's and the South's differing views of humankind's ecological interests and needs; see Chapter 4). The same is true for global warming, to some extent for acid rain and even for marine pollution. Rivers and water resources are typically shared between countries also, as with current concerns over access to and pollution of the Danube or the Jordan or the Rio Grande/Rio Bravo del Norte.

The cross-national character of the problems has meant that the respective countries' SMOs have seen opportunities for co-operation, opportunities which they have generally seized with greater alacrity than governments. Cynically expressed, governments have an interest in getting other nations' leaders to do as much about remedying a problem as possible so that they themselves have to undertake a minimum amount of work. SMOs do not obey this logic. By and large they are free to press for optimal action on environmental reform. Furthermore, they can use the green performance of the leading reforming country in any particular area of policy to berate their own governments. In the closing years of the last century, US legislation, on car exhausts and on freedom of information, was repeatedly pointed to by European campaigners with an argument such as, 'If the US government can insist on catalytic converters and still retain a car industry, surely European governments can do the same.' This line was adopted in the celebrated Greenpeace poster campaign, before the introduction of mandatory emission controls in Europe in the 1990s, which employed Ford's own slogan ('Ford gives you more') to point out that a British Ford did indeed give you more, in terms of harmful pollution, than an American Ford.

Lastly, the internationalism of the movement has been assisted because various international bodies, and in particular the European Union and the United Nations, have fastened onto environmental issues as a way in which they can augment their influence. The environment appeals to such bodies because they can argue that it is inherently international and because it can be presented as a public interest issue. If, for example, the EU is putting forward proposals which supposedly advance the public environmental good it is hard for national politicians to oppose these without seeming to argue out of pure self interest – seldom a rhetorician's favourite stance. Environmentalists may then find these international bodies afford a 'softer' lobbying target than do national governments. However, it should be noted that the costs and administrative demands of lobbying at this level tend to screen out smaller environmental organisations and thus to favour the larger campaign groups.

These factors explain why, in addition to the ideological reasons that environmentalists may have for favouring international collaboration, they have been successful in developing international solidarity. Of course, some practical features militate against this internationalism. Environmental SMOs develop a deep familiarity with their own countries' laws, politicians, civil servants and media. They are often so busy addressing these issues that the scope for internationalism is limited. Furthermore, when campaign organisations give priority to international issues this may lead them to neglect domestic topics and, in particular, to overlook the characteristic environmental problems of minority or disadvantaged communities within their own countries. Thus I am not suggesting that the green movement has transcended national barriers. But there are sociological and political reasons for believing that it stands a better chance of doing so than other putatively universal social movements; its only obvious 'rival' here is the women's movement which also appeals to supra-national objectives but with fewer pragmatic underpinnings for its universalist ambitions.

Finally, there is one further area in which the specificity of the ecological movement is apparent. This is the extent to which greens offer a critique of capitalism, an alternative value system and a view of the alternative society which they would wish to see ushered in. In other words, environmentalists offer a more comprehensive alternative than other NSMs; this is perhaps why they are often uppermost in the minds of analysts such as Melucci and why they seem to offer, as Touraine put it, a challenge to our great cultural orientations. The peace movement does not offer such a comprehensive and radical programme; it is difficult to argue that a commitment to peace necessarily leads to an antipathy to liberal capitalism.[2] Certainly, the peace movement does not of itself project an alternative blueprint for social organisation. The same would be true of the youth movement. Again, the feminist movement is the closest rival to the greens. But, to adopt the useful phrase which came into brief currency on the disintegration of the Eastern bloc, only the environmental movement offers a distinctive challenge to the idea that there is an 'end to history'. Greens have, as Dobson and others have argued (2000), a coherent political philosophy; they have distinctive views on the economy. They, of all social movements, have founded political parties in many countries, North and South, and experienced some electoral success. As Lowe and Rüdig presciently noted (1986, 537), 'Only the ecological movement represents a totally new political cleavage'.

However, the movement is not as coherent as this assessment implies. Just over a decade ago, and shortly before the UK Green Party turned in its best electoral performance (in the elections to the European Parliament) McCormick reported that the key supporters of the (UK) Green Party and of the leading environmental movements overlapped only to a limited degree (1991, 123–24). This finding was reinforced in a survey of British Green Party members which stated that 'only a minority of Green Party members have been involved in social movement activities of some kind to a significant degree' (Rüdig *et al.* 1991, 36). Furthermore, the leading British SMOs have been careful to distance themselves from the Green Party. The substantive focus of much of FoE's campaigning has been on suggesting improved policies for power generation, farming or transport. This advice has principally aimed at maintaining our standard of living while lessening pollution and resource depletion. Equally, they campaigned for the removal of CFCs from aerosols, not for the total banning of aerosols. Whatever the private convictions of leading campaigners, the demands of successful campaigning dictate that environmental SMOs wear a less than deep green garb. The in-principle transformational character of the green movement is thus inevitably diluted by the practical outlook of its leading SMOs.

A final aspect of this issue goes back to Berger's analysis of the knowledge class. As he himself notes, this class is in an ironical position of hostility to the very productive forces which have brought it into being. Without economic surplus this class would wither away. Thus, the very class-grouping in which environmentalism finds its most active and eloquent support has material interests antagonistic to a deep greening of society. The tensions inherent in the movement are replicated in its support base.

Conclusion

In this chapter, I have sought to explore the contribution of the sociology of social movements to our understanding of environmentalism and social change. I argued that only an essentially descriptive definition of 'social movements' is acceptable; none of the theoretically driven definitions succeed in specifying the characteristics of the movement. However, to accept a descriptive definition – an empirical generalisation – is no bad thing since the leading features of the environmental movement are well captured by this definition. Although there are many similarities between environmentalism and other movements, I identified three ways in which the ecological movement stands out: its intimate relationship to science, its practical claims to international solidarity and its ability to offer a concerted critique of, and alternative to, capitalist industrialism. According to Lowe and Rüdig (1986, 537) green politics has a peculiar inclusiveness and exclusiveness. It can accommodate other radical movements while resisting assimilation into their ideological agendas. I suggest that the three factors I have highlighted account in large part for this property. It should be noted though that the claims to universalism offered by the green movement and its critique of western society – however coherent in principle – are in practice limited. Leading environmentalists compete too much and distance themselves from deep green politics too much for the movement to have the practical coherence which inclusiveness and exclusiveness seem to imply; in this regard I judge that Lowe and Rüdig's claim is exaggerated.

3
Shell, a Sure Target for Global Environmental Campaigning?

with *John Forrester*

Introduction: globalisation and strategic options for the environmental movement

The environmental movement celebrates its global character. From the first 'Earth Day' to the Gaia hypothesis to the names which many environmental groups choose for themselves (EarthFirst! or Friends of the Earth (FoE)) there is a strong emphasis on equating the earth or the globe with the environmentalist project. Invoking the globe is inclusive; it embraces everyone, as well as all creatures, habitats and even the inorganic world.

In part, this global identity can be attributed to the notion that the environmental problems which the movement aims to combat are themselves global. Though they supposedly have 'global-ness' in common, environmental problems may be global in several different ways. In some cases this appellation appears justified because the problems themselves seem inherently global; global climate change is perhaps the most obvious case, though pollution of the oceans and the hazards of certain kinds of nuclear contamination would seem to qualify also. In other cases an environmental phenomenon is made global by its worldwide repetition; for example contaminated land or pollution from sewage or the majority of noxious pollution from motor vehicles. In such cases, the impact is local or regional, but international duplication makes the problem globally inescapable. Finally, some issues are rendered global by conceptual means; the transformation of concern for individual species (the panda or tiger and so on) into the worldwide loss of biodiversity is such a case.

However, this confidence that environmental problems are truly global and the associated assumption that environmental groups operate in the globe's interest can themselves lead to difficulties. In the first place, these factors can readily encourage environmental groups to be utopian in their assertions about global unity. If one can claim to act on behalf of the world as a whole, one might expect there to be no dissenters from the position adopted.

> It has clearly been in the interest of environmental philosophers and activists to claim that they are working for a global mission and that they represent the interests of the whole of humanity, perhaps the whole of the biosphere. But some of these claims are open to question. Even apparently global physical problems, such as the depletion of the ozone layer, are more severe in some areas (the poles) than in others. Claims about humankind's universal interest in solving environmental problems cannot necessarily be taken at face value. (Yearley 1995a, 215)

Secondly, the world's most powerful and well-resourced environmental groups, and the ones most successful in mobilising global imagery, are based in the 'North'. It has been suggested that they may accordingly contribute to the presentation of the North's view of environmental problems as the 'global' or as 'everyone's' position (see Bramble and Porter 1992, 350–51 and Yearley 1996a, 134–40). Indeed, the very assumption that we are 'all in one boat' might be contested by spokespersons for the South, who may deny that they share the same craft or, at least, argue that they are unwilling travellers on a Northern-crewed voyage (see Sachs 1994).

At the same time as these doubts about the genuinely global character of the environmentalist project beset the greens' embrace of global imagery and planetary rhetoric, there is an additional worrying possibility. In specific cases it appears that the very processes of cultural globalisation can be identified as the source of major environmental threats and harms. In particular, the World Trade Organisation (WTO) has been identified as a problem by environmentalists because of its propensity for ruling in favour of free trade and against the protection of markets or products on environmental grounds. The WTO favours liberalisation and has been inclined to view defences of trading policies in terms of environmental or humanitarian objectives as an excuse for protectionism. This worry is compounded by even more recent steps to harmonise internationally the rules for dealing with companies making transnational investments; transnational corporations (TNCs) are increasingly pressing to be able to take legal action against governments which appear to block their commercial 'rights', even on environmental or ethical grounds. In all, these steps are seen as evidence of global trends inimical to environmental protection.

From the point of view of Northern environmentalists and environmental movement organisations therefore, globalisation is double-edged. In devising a campaign and political strategy in a globalised context, they are faced with a number of tactical options. The majority of campaigns and social movement/NGO actions in the environmental arena are focused on particular problems such as biodiversity loss or the climate negotiations. But an alternative trajectory for global activism demands that one identify a global enemy. Although in some cases environmental problems arise from the acts of consumers or (as with the WTO) from the actions of policy-setting agencies, more often companies are the authors of environmental harms. It is accordingly tempting for movement organisations and activists to take global companies as the focus for campaigns and movement activities.

In the run-up to the Kyoto climate change negotiations in December 1997 a renewed emphasis was placed on the conduct of the world's leading oil companies. As Rowell notes:

> Technological advances and the maturity of existing oil fields have spurred oil companies to explore for oil and extract it from previously inaccessible or 'frontier' areas, both offshore and onshore. In many cases, such prospecting and production will have severe environmental impacts and serious social, ethical and cultural consequences. The challenge [for environmentalists, indeed for society] is not just to halt such exploration and extraction, but to halt oil consumption itself. (Rowell 1997, 99)

On this analysis, oil companies cause and encourage pollution locally and internationally; they favour the short-term use of irreplaceable natural resources; they demand transport infrastructures which frequently lead to major spills and ensuing despoliation of the marine environment, and their global pursuit of new reserves leads them into conflict with the interests of indigenous peoples and the guardians of relatively untouched habitats. They are therefore major and global antagonists (*not* friends) of the earth. Accordingly, the first objective of this chapter is to record the development of environmental protest against one particular oil company (Shell) as an indicative example of how such protests come about. The second is to reflect on the sociological implications for the environmental movement of strategies which focus around particular global (oil) companies.

Shell as a target for international environmental campaigning

Probably only Rio Tinto Zinc (RTZ) and, much more recently, Monsanto, compete with Shell as international icons of far-reaching environmental harms and as a focus for environmental protest. RTZ secured its position by the despoliation associated with its workings and by virtue of the repetitive environmental problems with which it was linked. Monsanto came to be regarded as the leading industrial proponent of agricultural biotechnology, particularly on account of its genetically engineered soya, cotton and rapeseed (canola). However, it is also notorious among environmentalists owing to its association with the generation of dioxins (particularly as contaminants in herbicide products) and the manufacture of PCBs (persistent and toxic fluids chiefly used in insulating electrical components). In late 1998 the campaigning British journal *The Ecologist* devoted almost an entire issue to critical articles on the company (1998). However, Shell appears to have overtaken RTZ both for the variety and thus 'depth' of the problems with which it is linked and because of the topicality of the problems with which its name is associated in the public consciousness. Moreover, it is more widely known and accessible than Monsanto and has, to date, been involved in a broader range of controversial policy arenas.

Though the catalogue of contentious issues is well known, it is worth recapping on them all briefly as a preliminary to subsequent discussion:

- Shell is associated with prospecting for oil in the seas off of East Timor. TAPOL, an Indonesian human rights campaign, held a protest meeting outside the official Shell Birthday Party on the company's centenary on 18 October 1997 (*The Ecologist Campaigns & News* 27 (6) 1997, C1).[1] Oil company interests in South-East Asia persist.
- In 1997 it was announced that Shell planned to start drilling for gas in an area of rainforest set aside by the Peruvian government as a homeland for nomadic Indian peoples. The project is forecast to last for some forty years. Of the ten trial drills reported to date, two have been in areas reserved for peoples with virtually no exposure to outside populations; early contacts with incomers are reported to have caused epidemics of whooping cough and serious influenza (Rowell 1997, 105). Though Shell maintains that it will use helicopters and

waterways for transport, thus avoiding the construction of roads which themselves can provide the means for further deforestation, Rowell asserts that: 'if Shell finds gas, however, a pipeline will have to be built which will entail building a road and which will in turn open up the forest for loggers and other colonisers' (1997, 105).

- Shell was, until 1998, a General Member of the Global Climate Coalition (GCC). Though 'general membership' is a second-class form of membership which limited the company's role in policy formation (*Corporate Watch* 4, June 1997, 13), the GCC's raison d'être is to lobby for the limitation of measures taken to lessen carbon dioxide (CO_2) emissions and to combat climate change. The GCC was said to be influential in encouraging President Clinton to take a reluctant attitude, towards far-reaching greenhouse gas emission targets at the Kyoto meeting. During 1997, FoE organised a letter-writing campaign to the head of Shell, UK, and had a twelve-foot shell 'demon' (with a demonic face and horns on the Shell 'shell') stalk petrol station forecourts. Shell followed BP (British Petroleum) in leaving the GCC when the oil companies recognised that majority opinion worldwide was shifting in favour of accepting the reality of humanly induced climate change. Association with the GCC was becoming more of a liability than an advantage.
- Shell also belongs to the World Business Council on Sustainable Development (WBCSD), an extension of the Business Council on Sustainable Development which was inaugurated in the run-up to the 1992 Rio Summit (Finger and Kilcoyne 1997, 141). The original BCSD had some fifty members and promoted a form of 'ecological modernisation', the pursuit of environmental improvement through technical innovation and changing business practices. Its first major publication, *Changing Course* (Schmidheiny 1992), offered a 'global business perspective on development and the environment', and included a list of case studies of good practice and business contributions to sustainable development. The Shell case study dealt with training and personnel development in its (now notorious) Nigerian operation (see below). WBCSD now responds to the UNCED agenda through the promotion of its 'overall conceptual framework, namely that the market offers the best solution for global environment and development problems, and that regulations are inefficient because they distort free trade' (Finger and Kilcoyne 1997, 141).
- Shell has become part of environmental campaigning folklore through the Brent Spar incident (see Chapter 9). The company had a 14,500 tonne oil-storage platform in the North Sea which had

reached the end of its useful life and accordingly needed to be disposed of. Despite their initial undertakings to bring the installation back on to land for dismantling, Shell came to the opinion that the Best Practicable Environmental Option (BPEO, a formal estimation of the best realistic course of action) was to strip the Spar as far as possible of contaminants and then to sink it in the Atlantic off the coast of Scotland. In 1995 Greenpeace spearheaded a well-focused campaign against this decision, involving an occupation of the platform itself, the provision of video footage of the mid-ocean struggles over the Spar, and a European boycott of Shell goods. Since consumers show little brand loyalty to particular varieties of petrol, the consumer boycott was very successful, particularly in northern Europe where sales (notably in Germany) reportedly fell by some 50 per cent. Greenpeace Germany provided the largest amount towards the costs of the campaign and German consumers were the most vociferous objectors: two service stations were firebombed while a further station was 'raked with bullets' (Thomas and Cook 1998, 7). Under pressure from Shell executives in Northern Continental Europe, the British wing of the company revoked its policy. Greenpeace celebrated its victory while the platform was towed to Norway for storage in a fjord. Only in 1998 were the revised plans for the Spar announced after a very widespread public consultation exercise: the platform was ultimately cut into sections and these sections were used as pillars to support a quay extension near Stavanger in the north of Norway.

- Possibly the most notorious case concerns Shell's involvement in Nigeria, specifically in Rivers State on the Niger delta. Shell has been accused of running its operations in an unusually ruthless manner, sharing few benefits of its economic success with local people, allowing repeated pollution incidents to contaminate river and coastal waters, treating the needs of local people with disdain in matters of siting its pipelines and other plant, and of flaring off gas in a way which is not only wasteful but also dangerous and polluting. In an area which is home to many local peoples, Shell has faced a series of protests during its decades of involvement with the Niger Delta. The most vociferous, associated with the Ogoni, led Shell to withdraw from much of its Ogoniland site in 1993. The antagonism was further heightened by the stance the company took over the imprisonment and eventual execution (in 1995) of a leader of MOSOP (the Movement for the Survival of the Ogoni People), Ken Saro-Wiwa. Many commentators believe that Shell could have exerted sufficient pressure on the government of Nigeria to have Saro-Wiwa's life spared. But during his final days, Shell took the line that it could not – as a mere foreign business – interfere in the country's internal legal and political processes. It is further pointed out by the company's critics that the viability of Shell's operations came to depend on protection supplied by the state paramilitary forces and that Shell appears to have armed certain of these forces, thus providing a means for implicating the company in the killing of civilians (Rowell 1995).

Shell is believed to be planning to resume operations in Ogoniland though it is also part of a partnership (with a 40 per cent holding) with plans to develop oil fields in southern Chad. In order to export the oil a pipeline will be needed, and this will run through Cameroon to the coast (Rowell 1997, 105). According to the Rainforest Action Network's 'Action Alert 135' of February 1998, the 'World Bank plans to fund an oil pipeline through Central African rainforests that will bring huge profits to Shell, Exxon and Elf while causing environmental havoc and threatening local populations – all with US taxpayers backing the deal. The oil companies are about to build a 600-mile pipeline from the Doba oil fields in Chad to coastal Cameroon, slashing through fragile rainforest that is home to the Baaka and Bakola peoples, communities of traditional hunters and gatherers. "Once construction begins, we'll see an uncontrollable influx of people in search of work – the result will be deforestation, wildlife poaching, and the loss of community land," says Environmental Defense Fund economist Korinna Horta'.

The Rainforest Action Network goes on to claim that:

> Observers fear that the project will create another Ogoniland, the Nigerian region devastated by decades of oil extraction and brutal military rule. Shell has worked with Nigeria's dictatorship to crush non-violent environmental organising among the Ogoni, who have watched more than 2,000 of their people killed during the last five years. Even as the plight of the Ogoni has come to public attention, the World Bank seems to have learned little from the oil conflict as it pursues funding the Chad-Cameroon pipeline in the neighboring countries. In Chad, the pipeline will likely escalate current conflicts as it traverses Doba, an area where the government is locked in a long-standing ethnic and regional struggle. In Cameroon, laying the underground pipeline will require moving into rainforest areas that are home to farmer and hunter-gatherer peoples, creating a climate for environmental [resistance].

Overall, therefore, Shell through its status as a global petroleum company would appear to qualify as a target for comprehensive environmental campaigning and as a rallying point for globalised pressure. In certain circles the company has become a watchword for a model of unscrupulous and environmentally harmful business. Thus John Vidal reported in *The Guardian* (14 March 1998, page 15) with some evident satisfaction that the Centre for Philosophical Studies at King's College, London, was having trouble attracting the most prestigious academics to its conferences because of its links with Shell (at the time Shell funded the centre to the tune of £60,000 annually). In the remainder of this paper our task is to examine the practicalities of social movement activities which have adopted Shell as a transnational target for globalised environmental campaigning.

Making a target out of Shell

This case study of anti-Shell protests has been based on campaign and associated literature from leading movement organisations, secondary sources and a limited number of group interviews with locally based campaigners in Britain and Ireland. Our analyses suggest to us that the globalisation of campaigning through the selection of a global target is faced with two principal sorts of difficulty.

Issues of orchestration

Shell is such a focus for campaign activity precisely because, in campaigners' eyes, it appears to be a transgressor on virtually all counts: it causes pollution in developing countries, fails to stand up for (or even offends against) human rights, injures the cause of indigenous peoples, opposes international environmental reform agreements and attempts to capture the international environmental agenda. But the very things which make Shell such a powerful stimulus to protest also offer practical impediments to successful campaign activity. Campaigning organisations have grown up with a sophisticated division of labour and with specialised expertise; in the UK those environmental organisations with the most sophisticated campaigning teams have advanced skills in particular fields but have, correspondingly, to limit their campaigning in fields where they are less qualified (Yearley 1993). To be adequately documented, Shell's 'offences' require the expertise of various environmental groups as well as Amnesty International, the World Council of Churches, Survival, the Rainforest Action Network and campaigners on trade. Except on extraordinary occasions, this would require a disproportionate investment of campaigners' time. At present the sense of an integrated campaign in the UK is maintained through the activities of *The Ecologist*'s 'Campaigns Section' which operates by referring readers to the website (or other information resource) of the appropriate NGO and through other similar *ad hoc* arrangements. In other words, precisely because of its diffuse organisational structure and the complexity of its political linkages Shell has no NGO 'mirror image'. A concerted, global anti-Shell or anti-oil-company campaign organisation has yet to develop.

Furthermore, not even the techniques for opposing Shell can necessarily be carried over from one arena to the next; the largely successful tactics adopted by Greenpeace to address the company's attitude to the Brent Spar were specific to that protest (see Bennie 1998 and the analysis in Chapter 9). As was mentioned above, the oil platform was out of the public eye; that isolation and the resulting fact that the Greenpeace boarding party could monopolise accounts of the conditions on the Spar was a peculiarity of the case. The international dimension of plans for marine disposal, particularly when media discussions of the Brent Spar often implied that the proposed dumping site was within the North Sea (as opposed to the Atlantic, west of Scotland), also gave environmentalists and publics outside the UK a vested interest in the outcome. Accordingly, campaign lessons may not be portable. SMOs compete and have to focus on specific targets. Most of these targets cannot be maintained indefinitely; in the end they have to move on.

This initial point relates to our second observation that the most conspicuous campaign achievements directed at Shell have come from specific restricted campaigns. In that sense, it appears significant that the Brent Spar campaign was clearly very successful but was also conducted in a limited way. It was dedicated to overturning one particular decision of Shell and thus could be treated (not least by Shell itself) as separate from the overall shape of company policy. Our supposition is that the kind of integration which an assault on a globalised company appears to offer does not sit easily with the proven campaigning strengths of SMOs.

These issues of orchestration also arise in relation to local activism. According to one informant, local movement activism was characterised to a large extent by individual participation:[2]

> I would say EarthFirst! [was my first point of information]...I think an interesting part of it was, how it was very much a lot of people, not groups, that came together.

This view was supported by another interviewee:

> I felt it was very much individuals, I mean you could identify people from different groups there, and presumably different groups networked. I mean the telephone network that I got part of at that particular time, I just literally rang everybody I knew who cared about justice.

Moreover, the groups cited as important to mobilising participants were sometimes atypical or unexpected ones:

Int: So it would be, for want of a better word, informal networks, it wouldn't be a recognised group.
R: Well, the most recognised group that I can recognise was my choir, a socialist choir...because that was one I could, you know, hit the button with quite fast, eh, and I knew people cared. That was probably sort of top of my list, after that it was just individuals, friends.

Given the informal nature of the protest, the question of co-ordination arises not only in relation to participation but also with regard to the activities selected:

Int: So I suppose the next logical question would be did you feel that there was anyone co-ordinating action against Shell here [locally]? Was there anyone telling people what to do or was it simply an agreement: this is what we ought to do. With the Greenpeace action, for example over Brent Spar, there was a very public face but there was also a lot of other things going on. Was there a sort of central co-ordination of action against Shell or was it simply groups of people who had this common interest and –
R1: I would have said the latter and I was very happy about that in the sense that that there was no, em, bureaucracy about it...
R2: Right, I think also it provided scope for individuals to present something in some way, such as the action which involved the mannequins, you know that was, like, you could hand out leaflets but make your protest in your own way, you think, those things that would have an impact.
R1: I think those protests never had anybody's badge on.
Int: How...?
R1: Until very late on when Survival kind of took it on because they had a particular tie-in, I mean that was not a take-over, I don't mean that...

R2: They could....

R1: They were plugging into something and they were using Survival's name and that was fine. But during the whole meat of the protest we never found people trying to badge it, that was good.

It appears that individual participation was highly valued. One of the respondents even talks about their communication to the public in individualised terms:

> most of the time it was direct appeal to passing traffic to hoot if they thought Shell should clean up its act in Nigeria. It was a very simple message.

Though this spontaneity is prized, without a co-ordinating body one runs the risk of the protest diminishing in an unplanned way:

> we had a very, very big turnout from which we had a presence at the garage every Monday morning at 8:00 the entire winter, and I was there I think just about every Monday morning through the winter. Em, numbers dwindled from that huge initial turnout obviously to three or four, but there was always a presence which was more or less taken on by Survival, and on particularly significant dates there were various things like the cardboard cut-outs, the nooses, so that there were visual props and stunts at various times.

Additionally, there are limits on the kinds of activities which can be undertaken in a 'spontaneous' way. When asked about local activities, participants responded as follows:

Int: We briefly emm mentioned what we did, which was essentially a public, visible protest. In front of Shell filling stations, were those the only tactics you used, did you do anything else?

R1: And the public meeting!

Int: And the public meeting, sorry.

R2: And there was letter writing

Int: And there was letter writing as well to –

R2: To the Nigerian government

Int: To the Nigerian government

R2: and to the English government.

The range of activities they were able to engage in without the support of bureaucratic SMOs was plainly limited.

In this section we have seen that, though Shell serves to coalesce and concentrate a great many environmental problems, the company does not necessarily offer itself as a ready or straightforward target for globalised opposition. Campaigners, at both local and national levels, face significant problems of co-ordination.

Theorisation

Our second point is that generalised opposition to Shell (indeed to the oil industry) faces difficulties because opposition does not appear to be based on a consistent ideology or analytical understanding. For some campaigners, for instance, Shell's involvement in Nigeria and the plight of the Ogoni now appear to stand as a self-explanatory paradigm of environmental and socio-economic wrongdoing. The iconic status of this example seems almost to mean that the specifics of the case need no further examination or explanation. The rhetoric of the Ogoniland suffering has become an identifiable and easy shorthand for problems ensuing from TNCs' involvement. As noted above, the Rainforest Action Network identified people's 'fear that the project will create another Ogoniland' in southern Chad. Rowell (1997, 105) uses the same wording to comment on the Chadian case even though the colonial history, the ethnic characteristics of the divisions within local society, and a range of other geographical and technical factors may differ.

EarthFirst! (in its *Action Update No. 41*, July and August 1997, 1) observed the need to tackle a more theoretically integrated target, and called for '100 days of action against fossil fuels'. The text goes on to explain that:

> A coalition of anti-oil campaigners are calling on activists to strike at oil targets over the next 3 months in a concerted 100 days of action. The message is simple: we need an end to the dependency on fossil fuels. Activists involved with RTS [Reclaim the Streets], Delta, Corporate Watch, Platform, 90% Crude, Greenpeace, Oilwatch and Cardigan Bay EF! and others will hold the oil industry accountable for their crimes. The hundred days will begin on August 23rd [1997] and end with the Kyoto Climate Summit, where world leaders may begin limiting CO_2 emissions but will fail to tackle the real cause of climate change – our mad quest for oil. Up until now, anti-oil action in the UK has been either in disconnected bursts or focused on the effects of oil: Shell in Nigeria, the Brent Spar, exploration in Cardigan Bay, BP in the Atlantic Frontier, Mobil in Peru, Milford Haven, The Braer and Sea Empress Disasters,[3] pollution, workers' rights, oil spills, petrochemical toxins.

> Now with the threat of rapid climate change we must end it before it ends us. The oil industry is large but cumbersome and has never before dealt with an all out assault on its right to exist!

Even here, the key charge against contemporary patterns of consumption trades on the identification of 'our mad quest for oil'; yet the meaning of this central phrase is left unexplicated. Without some analytical understanding of the source of our 'fixation' on oil it is very difficult to determine which course of campaigning action to adopt. Anti-Shell campaigners devote little public attention to addressing this problem, at least in part – one assumes – because it is easier to agree that oil companies are in the wrong than to secure agreement about what precisely is wrong with Shell or the industry as a whole. The pervasiveness of the oil companies and their products was discussed by one respondent:

> Well so maybe it boils down to plastic bags? Making people aware of the connection between what happened to Ken Saro-Wiwa and that plastic carrier bag you've just picked up in the supermarket. [At the] basic level, because even people who said, okay I'll use my car less, I'll think about, I mean that [aspect] of culture has come into their consciousness, will still pick up six plastic bags at Waitrose, Sainsbury's – that hasn't, that was the message that we actually finished the public meeting, that it's as simple as that plastic bag.

This observation appears to acknowledge that the oil industry is only being challenged in its most obvious manifestations.

Associated with the problem of 'theorising' the opposition comes the question of how to judge the success of one's ventures. When local activists were asked about this, they answered:

> We felt connected with people in other parts of the world, and it is my observation that we didn't ever try to blockade the [Shell petrol station] forecourt. People were free to drive in and they did. That wasn't, the purpose was not to blockade, it was to raise awareness, and one of the great joys for me was that a coach of schoolchildren came by every week while we were demonstrating and they were all shouting away and they were looking for us, cheering, and I thought they've seen us week after week, they'll be asking questions about that's, that is a success in itself.

Responses of this sort are commonly tied to a defence of the value of action in terms of the worth of political action per se. In McKay's work on informal environmental protest, such a view is succinctly espoused by 'John' of the radical political theory magazine *Aufheben*:

> [B]y adopting direct action as a form of politics, we . . . look to ourselves as a source of change . . . Therefore the key to the political significance of the . . . campaign lies less in the immediate aims of stopping the road [or whichever issue is the focus of the protest] and in the immediate costs we have incurred for capital and the state (although these are great achievements and great encouragement to others), and more in our *creation of a climate of autonomy, disobedience and resistance* . . . Thus, this life of permanent struggle is simultaneously a *negative* act (stopping the road etc.) and a *positive pointer* to the kind of social relation that could be (quoted in McKay 1996, 127, his italics).

This communal and community action helps protesters feel empowered in the face of globality and its success cannot be gauged simply by whether the protesters' preferred outcome is followed or not. Further, the rationale behind 'corporate' protests such as that orchestrated by single-issue groups like Greenpeace over Brent Spar is fundamentally different from the protest against the corporate, global lifestyle which is engendered by TNC.

Concluding discussion

On the face of it, Shell offers an attractive, global target for campaign activity. It offends broadly and, given its global reach, it offends widely. One might suppose therefore that global campaigning would target a global quarry. Through its public commitment to environmental causes (for instance its 'Better Britain Campaign' and its recent World Wide Fund (WWF) award for environmental reform in Canada) Shell is clearly 'asking' to be judged by lofty standards. But it appears to us that there are sociological reasons for scepticism about the ability of a global target to nucleate sustained global protest or campaigning. The target is too diverse, too unwieldy and demanding of too great a variety of expertise to be readily amenable to this approach. Ironically, therefore, aside from their rhetorical value, global environmental 'sinners' may not be the unifying opportunity for global environmental action that one might have anticipated. The 'struggles' against global enemies may become iconic, but it is hard to press the advantage home in terms of achieving comprehensive change or reform. Campaigners have, however, in recent years identified other kinds of institutional offenders: notably the WTO, the World Bank and IMF, and meetings of the political leaders of major capitalist countries (for example the G7 and G8 meetings).[4] We suspect that these protests face similar practical obstacles although research on that topic remains to be undertaken.

4
How Environmental Problems Come to be 'Global': Sociological Perspectives on the Globalisation of the Environment

Introduction

On the face of it, no issues are better suited to treatment in terms of globalisation than environmental ones since leading environmental threats appear physically or biologically global. There is only one Earth, only one, interconnecting biosphere. During the late 1980s and 1990s, the prime focus in international environmental problems settled on successive global issues, first on ozone depletion and on the atmospheric circulation of POPs (persistent organic pollutants), and then on biodiversity protection and global climate change. The label of 'global environmental issues' came to feature prominently in research agendas and environmental action programmes throughout the Northern world. At the same time, these countries' inhabitants have repeatedly been offered new identities as citizens of planet Earth. Novel imagery has proposed that the world's population is the crew of spaceship Earth. Voluntary organisations have offered us the chance to become *Friends* of the *Earth* (FoE) or supporters of the *World Wide* Fund for Nature. We are invited to put the *EarthFirst*! These verbal images have been reinforced with innumerable pictorial variations on the Earth seen from space: Planet Earth, a brilliant blue and green jewel, hangs isolated in the vastness of the Universe. And this global identification seems to fit precisely with the twofold notion of globalisation proposed by Robertson. As he expresses it: 'Globalisation as a concept refers both to the compression of the world and the intensification of consciousness of the world as a whole' (1992, 8). The concept directs analysts both to the actual global ties which bind people or societies together, and to the subjective appreciation of the importance of those ties.

In the first place, then, environmental arguments give us specific grounds for viewing the world as – in fact – an evermore interconnected place. These connections can broaden environmental problems to the global level. Ozone-depleting pollutants manufactured primarily in the North and released by (some) consumers almost the world over threatened to thin the world's protective ozone layer, especially at the poles. The world's oceans are minutely contaminated by oil, oil extracted at very specific locations across the globe but distributed by industrial processes and transport so that all seas are now contaminated. Proposals to stock fish farms with super-fast growing, genetically modified (GM) salmon have been opposed on the grounds that escapees would contaminate the genetic stock over thousands of kilometres, thanks to the vast distances travelled by wild fish. While the complexity of ecological connections often helps ensure that pollutants may be spread the world over, they can also be concentrated unpredictably. Many pesticides and also much radioactivity are subject to concentration by biological systems so that emissions, rather than being dispersed, can accumulate. In this way, global processes may manifest themselves as acute local difficulties with no obvious local cause. In the second place, these physical processes are not just an objective enactment of globalisation. We are invited to use these objective connections as an underpinning for subjective appreciations of global oneness and identification. The Earth is our common habitat and – in some sense – we all have a stake in its protection. The idea of global environmental issues exemplifies this common interest since, almost by definition, everyone would benefit from combating problems that are truly global. Indeed, so effective has this identification become that, in practice, global discourse has become synonymous with the environment. When 'Earth Day' is celebrated, it is clear that 'the Earth' is understood through the lens of environmentalism. Earth Day is not a day dedicated to celebrating the earth (soil) under our feet or the fact that we live on the third rock from the sun; it focuses on the planet as an environment for life.

In this chapter I propose to consider the senses in which 'global' problems are global, the ways in which their global character is constructed and contested, and (using examples from international agreements) the extent to which this global character makes them subject to common responses.

A canonical global environmental problem

It is now almost routinely accepted in official statements, in regular news coverage and by many campaigners that the major environmental problems are 'global'. In particular, climate change is the canonical global environmental problem. Carbon dioxide (CO_2) gas – the principal greenhouse gas – is invariably emitted by all the people of the world, if only just through breathing, but more dramatically insofar as they burn fossil fuels or wood. But the greenhouse gases each of us emits do not affect us alone. The CO_2 molecules we liberate stay in the atmosphere on average for around six years (Silvertown 1990, 78–82), meaning that they drift around the globe. Inevitably, our individual emissions get mixed in with everyone else's. Thus, the greenhouse phenomenon, where the Earth's atmosphere retains more heat than it formerly did because of the presence of additional insulating gases, appears to operate at a planetary level. Reducing my emissions is not generally likely to lessen the impact of global warming or climate change at my locale any more than at any other specific point on the globe. Furthermore, atmospheric warming is thought likely to induce physical and biological changes at a planetary level. For example, a warmer biosphere will lead to an expansion of the oceans. Sea-level rise, perhaps by tens of centimetres, will occur all over the world's oceans. Similarly, changes in the weather system will have implications for climate across the whole world. We are all implicated in the cause. We are all likely to be affected by the consequence. And we cannot isolate our individual responsibility for the pollution because physical and biological systems lead to connections taking place across the planet. No polluter is an island.

Convincing though this account is, we need to examine whether other environmental problems are global in all the same senses and, if they are not, to consider how the epithet global has come to be so ubiquitous. It is also necessary to ask whether we 'all' have the same stake in resolving such global problems. I suggest there are three conceptual dimensions to these issues: the supposed global character of the phenomenon; the conceptualisation of the issue as global; and the deployment of the label 'global' in environmental policy making and campaigning.

A problematically global environmental problem

At the opening of the twenty-first century the only environmental issue to rival the international prominence of global warming was the spread and regulation of genetically modified organisms (GMOs), especially genetically engineered foodcrops. There were manifold issues at stake in this context, but to simplify one can state that US and some European firms had begun to produce GM crops with two principal forms of putative benefit: some were engineered to have in-built resistance to pests while others had genetic tolerance towards particular weedkillers. In the latter case, the intended benefit is that fields of crops can be treated with the herbicide during the growing season and everything but the crop would be killed off, making weedkilling more thorough and economical. Prior to GM, farmers could not act in this way since, alongside the weeds, the weedkillers would also wipe out the crop. The crops initially targeted for these two treatments included soya, maize (corn), rape (canola) and fodder beet. These crops are traded internationally and are often used in animal feeds, with the animals themselves (dead or alive) being further subject to international marketing. Campaigners against GMOs claimed that these crops had not been proven safe for people to eat. With perhaps greater credibility they also claimed that there are specific environmental drawbacks associated with the cultivation of these forms of GM crops.

Possible environmental harms are associated with both the strategies for biotechnological progress mentioned above. On the one hand it is feared that herbicide-resistant crops will transform fields into 'wildlife deserts' where only the crop plant remains standing. Birds and insects that depend on anything apart from the crop plant would not survive in such fields. Even field margins (often important as wildlife refuges) would be susceptible to spraying. There are further fears that the herbicide-resistance trait could be passed to related crops, giving rise to 'super weeds' impervious to weedkillers. In the case of genetic resistance to pests, there have been two prime concerns. Many commentators have observed that this procedure could promote the evolution of tolerant pests able to cope with the biological control measures. The second concern is that species other than the target pests might be affected. Experimental evidence, from laboratory simulations, suggested that biological insect resistance being bred into maize (corn) could be distributed in the environment by corn pollen falling onto neighbouring plants (see Pimentel and Raven 2000). In the case in question, it was suggested (perhaps rather implausibly, see Chapter 10) that monarch butterflies – celebrity insects of North America – could have their larval stages endangered by insect-harming elements in maize pollen blown onto their favoured foodstuffs.

In this case, one has a product where the emphasis is very local, at the level of the individual soya plant or the field of beet. But the possible ramifications are extremely widespread. Successful seeds which seemed to farmers or food companies to confer significant advantages could become widely diffused. Within two or three years – such would be the hope of seed companies – a single variety could come to be cultivated in

North America, Europe, South America, Australasia and possibly the Indian sub-continent. Any resulting environmental problems would thus be distributed over many continents even though this would arise through the repeated application of a local technology. But the issue has other supranational aspects too. For example, the case of the USA's monarch butterflies could putatively affect Mexico also since the adult butterflies migrate there. Furthermore, if weedkiller-resistance developed in wild crops or in escaped cultivars, it could spread as a trait across national boundaries. Similarly, genetic modification of fish, farmed salmon, carp and tilapia, could give rise to interbreeding and genetic pollution spreading internationally through rivers and watercourses.

The issue can be globalised through trade and associated campaigning in another sense also. For example, initial enthusiasm for the cultivation of GM crops in North America was dampened by resistance from European consumers. Given that, at the time of planting, farmers had to judge the state of the market after harvest, American growers were torn between sowing some GM crops (which appeared to them as farmers advantageous) and some non-GM crops for the export market but having to keep the two separate *or* growing only non-GM produce. Either way, profits were likely to be reduced by the costs of keeping the two crop-types separated or by the likely diminution of yield from not adopting the GM technology. Frustrated by the difficulties experienced by their farmers in gaining guaranteed access to the European market, the US authorities have considered bringing this issue before the World Trade Organization (WTO). The situation was made more complex by campaigners' increasing interest in tracking GM items back up the food chain. Though European (and other) consumers had become concerned about the likelihood that they themselves were eating GM produce, campaigners tried to broaden the matter by focusing on the probability that consumers were eating livestock which had itself eaten GM rations.

Finally, seed companies tended to present these genetic modifications as only the start of the latest revolution in agriculture. They were after additional characteristics which could enhance their seed products: not simply ones which appeared to aid in the routine cultivation process, but genetic additions which might confer medicinal properties on the food or assist with cultivation under adverse (drought-prone for example) conditions. The search was on for genetic material from organisms that manifested such desirable characteristics. And given that the greatest biodiversity reserves are in the developing world, this meant that Northern companies have been prospecting for genetic resources in the South. This has led to concerns about the allocation of ownership of rights over this genetic material and to accusations of genetic colonialism.

Thus, it is evident that the GM issue too has many transnational dimensions: the threatened spread of genetic contamination, intellectual property rights, trade and the involvement of the WTO, and the international impact on wildlife. As with global warming, it is hard for countries or even consumers to insulate themselves from the issue. And, in case one is tempted to argue that the problem is not inherently global in the same sense as climate change and global warming, the international spread of GM crop strains or the inadvertent dispersal of GM micro-organisms might turn out to be globally ubiquitous. The hazards of GMOs are also potentially global in nearly every other sense. But one does not see the same rush of political authorities to describe the risk of genetic pollution as a leading global environmental problem. This, I suggest, leads us to two conclusions. First, environmental problems do not qualify to be 'global' simply by meeting physical, ecological or biological criteria which distinguish global from less-than-global ones. Second, it unsurprisingly suggests that what wins the appellation of a *global* environmental problem is not so much the inherent characteristics or indeed the severity of the problem, but the outcome of a politics of problem-labelling.

Attaching the global label

It is one thing to propose that a key issue is the labelling of environmental problems as global, and a rather more complex one to identify the processes by which that labelling occurs. The major force in labelling has come to be policy agencies in the North. Following the Rio Earth Summit in 1992, Northern officials employed the global terminology to refer to issues which they regarded as urgent and which they wanted everyone to play a role in addressing. As has been widely observed, the common commitment to an ideal of 'sustainable development' concealed tensions between the demands of the South for much more development (to combat poverty and indebtedness) and the emphasis of the North on higher environmental standards. Many argued that sustainable development was oxymoronic. The conceptualisation of issues as global environmental problems operated in a similar manner to 'finesse' North–South differences. It provided a legitimation for the North encouraging, even enforcing, environmental measures in the South while allowing the North to classify other issues (for example, local industrial pollution) as comparatively unimportant since they are not global.

Let us take the example of global warming. Configuring this as a matter for globally co-ordinated action justifies the North – for the globe's sake – taking an interest in the fate of Brazil's (and other Latin American, West African and South-East Asian nations') rainforests. Forest management in Europe and in North America is still principally seen as a national matter. But the alleged global significance of the Amazonian forest, the 'lungs' of the planet, means that this habitat can be presented not as a matter for national management but as everyone's joint resource and responsibility. This is not to take a sceptical view of the arguments about deforestation and global warming, nor to argue for the wisdom of national policies in all South American, West African and South-East Asian states. But it is to suggest two things: first that, typically, the global label is applied unevenly – to Amazonian forests but not to the mass consumption of off-road vehicles and sports utility vehicles (SUVs) in the North – and second that it provides Northern countries with an apparently principled basis for intervening in some other nation's policies.

Furthermore, to present an issue as global is often tantamount to implying that it is equally – and highly – urgent for all. If a problem is serious enough to be a global-level worry, that might seem to indicate that we should all worry about it. But, even accepting that global climate change is indeed global, that does not demonstrate that it is equally urgent everywhere. By and large, the policy discourse about climate change focuses on the problems it may cause for coastal cities, for flood-defence, for agriculture and so on. It addresses the probable costs and socio-economic impacts of the changes that are believed likely to arise. From the perspective of a poor, slowly developing country these costs may seem low and remote compared to the costs of other forms of pollution or the environmental consequences of poverty. It is for approximately this reason that the Kyoto Protocol on limiting emissions of greenhouse gases was agreed only to extend to the industrialised world (at least, in the first instance). Leading developing countries saw no reason to sign up until the industrialised world had demonstrated its bona fides. The announcement in spring 2001 by President George W. Bush that the USA would not ratify the Kyoto agreement was legitimated in large part by arguing that, unless everyone (including China and India) was committed to greenhouse reductions, the costs would fall unevenly on those volunteering to make the cuts. US self-interest dictated that the only commitments to which the USA could bind itself would be truly global ones.

Up to this point I have largely argued that the politics of 'the global' turn on decisions about, and the treatment of, those issues which are to be included. Making tropical deforestation a matter of global concern supplies an apparently disinterested motivation for broadening the scope of the international community's interest in the domestic politics of the South. In a directly analogous manner, when the World Bank became involved in the management of the GEF (the Global Environmental Facility, a scheme of loans for the protection of environmental quality), the loans were only available for specified global issues, for example biodiversity conservation and actions designed to restrict greenhouse gas emissions. Applicants could not negotiate over what 'global' meant; only specific policy arenas were deemed global. But this suggests that, as well as specifying what is included within the global label, there is the complementary question of *exclusion*. Lack of access to elementary education or even worries about contamination through the export of GM foodcrops do not count as global environmental problems since this label is denied to them by the most influential classificatory agencies.

While the actions and interpretations of official policy analysts have had a large impact on the use of the 'global' label, environmental and conservation NGOs have played a significant part also. During the 1970s and 1980s the growing environmental movement played a largely oppositional role, campaigning against nuclear energy programmes, industrial dumping and air pollution. By the 1990s they too were attracted to international environmental campaigning. Many problems appeared to exceed national barriers and it was becoming clear that attempts to improve the environment of the North could result in the export of pollution to the developing world. It seemed unethical to campaign to promote one's own quality of life if that was achieved at the cost of other nations' declining environmental well-being. Already ideologically disposed towards the idea of environmental issues as everyone's problem, NGOs readily adopted the language of global environmental problems. Furthermore, if problems were global then that gave campaigners a viable mandate for operating outside the countries in which their supporter base and fund-raising lay. They were also able to exert moral pressure on domestic governments by stressing those governments' international obligations. And in the matters of biodiversity conservation, the protection of the ozone layer and the countering of global warming they were well served by a straightforward conception of the global interest. When in the early 1990s a US-based environmental lobby group, the World Resources Institute (WRI) tried to drive forward campaigning on global warming by naming the world's biggest greenhouse polluters, they numbered five developing countries among the top ten offenders. They did not anticipate that developing-world environmentalists would criticise their report on the grounds that it took the global framework for granted, failing to distinguish between countries such as India and China which (arguably at least) emitted a lot of unavoidable greenhouse gases and – on the other hand – Northern states (the USA and Canada in particular) which had wantonly cultivated high-carbon lifestyles. In the view of critics such as Agarwal and Narain (1991), the WRI's simplistic adoption of a globalist outlook had led the organisation to overlook major differences between nations.

I have now illustrated ways in which the global label has been applied by both policy makers and by NGOs. But it still might be argued that the epithet 'global' is not simply a label precisely because it can be scientifically underwritten. On this view, some processes are – in fact – global while others are not. Indeed, the language of science has also been decisive in the development of the discourse of global environmental problems and it is to that issue that I now turn.

Scientific generalisation and conceptual globalisation

In his work on globalisation, Robertson emphasises that he is not solely concerned with the 'objective' aspects of the process; he stresses the independence of the cultural sphere. He and Lechner assert that 'the thematisation of globality is presently the object of symbolic constructions' (1985, 108) and that there are '*independent* dynamics of global culture' (1985, 103, my italics). This sets them apart from authors who view globalisation as the consequence of economic harmonisation alone. In place of a single theory about the cause(s) of globalisation, Robertson's emphasis is on the heterogeneous elements promoting globality; this leads him to develop a historical model of stages of globalisation (1992, 58–60). What I want to suggest now is that the universalising discourse of science is a neglected part of the cultural processes sketchily indicated by Robertson.

In the late 1960s and early 1970s, Habermas and other 'Critical Theory' authors outlined a critique of the role of scientific and technical decision-making procedures in Western states. In brief, their argument was that popular participation in politics was declining as more and more decisions were taken in line with the dictates of alleged scientific and technical rationality. By invoking the 'rationality' of certain courses of actions, governments could free themselves from democratic accountability since, supposedly:

> The solution of technical problems is not dependent on public discussion. Rather, public discussion could render problematic the framework within which the tasks of government action present themselves as technical ones. Therefore the new politics of state interventionism requires a depoliticization of the mass of the population (1971, 103–104).

Habermas referred to this process as the scientification of politics. In concrete terms, it raised the possibility of the usurpation of political power by technocrats, since 'The state seems forced to abandon the substance of power in favour of an efficient way of applying available techniques in the framework of strategies that are objectively called for.... It becomes... the organ of thoroughly rational administration' (1971, 64). Scientification is accordingly a threat to democratic control over politics and policy.

From the perspective of the early twenty-first century, it is difficult to identify directly with Habermas' anxieties. He foresaw a growth in government intervention, legitimated by technocratic rationales – while the current trend is towards a shrinking of the state, this shrinking itself justified in terms of the 'logic' of the global economy. However, his point about the potential opposition between scientification and participatory or directly accountable democracy is still useful and provides a way of talking about the practical weaknesses in the application of the universalising discourse of science to global environmental problems. It seems that in this case what Habermas feared of the state has, in significant respects, become true of the international policy community.

Habermas was anxious about the growth of technocratic management because it allowed the state to get away without defending its policies to its citizens. It gave the state the chance to say that a policy had been decided on because the relevant experts had deemed it technically necessary. In the case of the negotiations leading up to the international treaty on ozone depletion a similar process operated (Yearley 1996a). Politicians, scientists and lobbyists in the North believed that ozone depletion was becoming a global problem. Since it was a technical problem, it was assumed that scientists from any country would have agreed about the problem though in fact virtually all the concerned scientists, all the detection equipment and all the major negotiators were from the North. Scientists and environmentalists had few reservations about representing the interests of the globe because this was a technical problem to which there were rationally calculable, technical solutions.

Given that they were operating in the interests of the globe on the advice of scientific experts, Northern policy makers were happy to impose their solution on developing nations, employing the threat of trade sanctions to win compliance (for details of the negotiations see Benedick 1991).

However, there were assumptions made in the technical framing of the problem which Third-World technical delegates – had the Third World been more broadly represented – might have questioned. These could, for example, have included the small range of policy alternatives considered. The emphasis was on finding replacement chemicals to play the role of ozone-depleting substances in cooling equipment, rather than on questioning whether building designs which required air conditioning should not themselves be changed. There was also widespread concern about the allocation of intellectual property rights over the new CFC-substitutes. It appeared that the use of the new CFC-substitutes could become mandatory without Southern countries being granted any assistance in paying for these chemicals which were far more expensive than CFCs had become. The conviction that science speaks objectively and disinterestedly seems to mean that one need have no qualms about excluding other people from decision-making since they would, in any event, have arrived at the same conclusions as oneself.

I suggest that this approach can be summarised in terms of the following table (Table 4.1) which addresses the assumptions behind allowing scientific expertise to speak of the global need.

The limitations of scientific expertise in resolving global environmental disputes are ironically highlighted by the GMO case discussed earlier. As mentioned above, US companies have urged that European resistance to GM imports should be combated by appeals to the WTO. Their argument is that there is no scientific evidence of harm arising from GM food and crops, since these products have all passed stringent regulatory hurdles in the US system. On this view, labelling of GM produce in the European market (the procedure favoured in most European Union member states)

Table 4.1 The 'logic' of the universalistic discourse of science

- There is a common (scientific) currency for assessing the global extent of a problem
- This currency is not arbitrary but well founded
- Any properly trained scientist would be able to use it, and it would be recognised as valid by other scientists
- Thus, any properly trained scientist can stand in for any other
- 'Democratic' representativeness, or representation from 'stakeholder' nations or regions, is not therefore required.

is discriminatory and an unfair trading practice since it draws consumers' attention to an aspect of the product which has no relation to its safety. The label 'warns' the customer of the GM content but, if that content is not dangerous, then all the label will do is penalise US and other GM-using suppliers. According to this way of seeing things, the WTO should outlaw this labelling practice as an unjustified impediment to trade. European consumer advocates argue, by contrast, that the US-testing has not been precautionary enough and that properly scientific tests would require much more time and more diverse examinations than have been applied in routine US trials. The prime difficulty of course is that there is no further scientific body to which to refer these competing claims and so the tacit assumptions of the scientistic approach breakdown. By and large the US and European scientific establishments support their respective governments' positions and there is nowhere else to turn. This conflict may turn out to be the first of many cases where the WTO's judgement will turn on disputed scientific advice.

Two conclusions follow from these points. First, while the discourse of science is supposed to offer objectivity and disinterested authority, in practice the application of this discourse to global environmental problems does not resolve the issues once and for all; it can itself even give rise to accusations of partisanship. Second, international agreements, when they are reached (and even when they deal with matters, such as ozone depletion, apparently amenable to scientific analysis), will generally not arise from the absence of conceivable scientific alternatives, but from the parties' willingness to suspend argument – perhaps because there are political or economic incentives to strike a bargain or because a country lacks the resources with which to continue to make the argument.

Concluding discussion

The question with which this chapter began was how it is that environmental problems get to be 'global'. The answer proposed here is that it is significantly a matter of labelling and, so to speak, social construction. Though one can make an apparently persuasive case for the physical, 'natural' global-ness of some global environmental problems (such as global climate change), other candidate problems with similar 'credentials' are denied this appellation. In this chapter, I have reviewed some of the social processes involved in this labelling. Some operate at the level of official policy processes; others are less straightforward and involve, for example, the campaign strategies of Northern NGOs.

Of particular interest is the role of scientific discourses in appearing to underwrite the identification of specific problems as global. If problems' global status could be vouchsafed by science, then that would seem to diminish or perhaps eradicate the scope for social-labelling processes. However, I have argued that, both in practice and in principle, the scientific identifying of 'global' issues is itself open to the influence of labelling. Not all parties have equal access to the framing of the scientific questions; not all countries have similar scientific capabilities in order to construe the problem in the way that seems most convincing from their own perspective. And even in some disputes between northern environmental policy makers, scientific institutions cannot resolve the parties' differences.

To say that environmental problems get to be 'global' through labelling is, of course, not to say that many environmental problems are not of compelling global importance. It is, however, to invite us to question the repeated identification of the 'usual suspects' in international environmental policy line-ups.

Part II

Studies of Environment, Law and Public Policy

5

Bog Standards: Contesting Conservation Value at a Public Inquiry

Introduction

Decisions about the environment get made in a wide variety of ways. Some are made by consumers, others by companies, and many large ones by governments and regulatory agencies. Lots of decisions pass without comment, without even being noticed. But a few become explicitly debated. Some become the focus of highly public controversies (as in the case of the Brent Spar, see Chapter 9) while others are handled through officially sanctioned mechanisms for the resolution of disputes. The case study that forms the focus for this chapter is an example of this last category.[1]

This case concerns a public inquiry into the conservation value of an area of lowland wetland, a peat bog, in Northern Ireland. As described below, the inquiry was run as a formal legal hearing. Part of the evidence brought forward in this hearing came as testimony from naturalists. Scientific witnesses facing cross-examination have to confront challenges arising from the need to have their arguments publicly understood by lay persons, the assessor or judge and, if there is one, the jury. In this context, as exemplified in a redoubtable paper by Oteri *et al.* (1982), scientists can be made to suffer under public scrutiny. The performance of scientific witnesses offers an opportunity for the analysis of a 'naturally occurring' exercise in the how well experts stand up to public inspection. Of course, the legal context should not be taken as a representative instance of public testimony. For one thing, as will be seen, there are elaborate rules of legal reasoning and procedure. While these may derive from everyday notions of fairness and proof, they differ considerably from common-sense reasoning. Scientific arguments may well be treated very differently in court from the way they are ordinarily handled by members of the public. None the less, it is the task of the court to allow lay persons (jurors, assessors and so on) to pass judgement on the basis of the available evidence. Some sort of public understanding of science (PUS) must therefore (be seen to) have been achieved. The elaborate legal procedures allow this achievement to be studied in close detail and it is this feature that I propose to take advantage of in this chapter.

In their paper, Oteri *et al.* essentially provide a series of interrogational formulae to be employed when examining chemical analysts in drugs trials, typically when acting for defendants. While chemists' evidence on the identity of, say, heroin is routinely accepted in court, the authors maintain that it would frequently be possible for the defence to undermine the expert's testimony if such an attempt were made. They view defence lawyers as unnecessarily defeatist in their attitudes. The assault on scientific testimony may be conducted in a number of ways. The factual identification may be challenged, for example, by suggesting or implying that some other substance might have been mistaken for heroin during the analyses or that traces of the drug were present in the equipment from previous tests. On the other hand, the analyst himself or herself may be the target of the questioning: had they ever taken a course in drug chemistry? were they really up to date with the latest testing equipment? and so on. The moral charge attaching to drugs can also be called into play. If the analyst is not very familiar with the substances then his or her experience is in doubt; if he or she claims great familiarity with drugs, this piece of evidence may be developed in such a way as to imply that their experience of drugs is not wholly professional.

The detailed 'recipes' that Oteri, Weinberg and Pinales offer display an interesting combination of what we may at this stage refer to as understanding and misunderstanding of science. They advise that the 'examiner should learn all the possible tests that could be used for the analysis of any particular drug' (1982, 253). They also explain the importance of asking about control tests (to make sure that heroin did in fact give the anticipated reading on that equipment that day) and blank tests (to ensure that a false positive was not obtained). This requires that examiners go a long way into the technical culture of the analyst. Yet the authors advocate that this technical familiarity be deployed for adversarial advantage too. In doing so they appear to encourage a wilful failure of understanding. For example, they recommend drawing on the in-principle unverifiability of all scientific knowledge to elicit the reply that the analyst is not 100 per cent sure of the identification of, say, heroin; scientific experts will typically agree that, as scientists, they cannot be 100 per cent sure of anything. This admission can then be developed into a practical doubt about the particular identification in hand (1982, 254).

Since their paper deals with legal advice and is not a sociological account of how expert testimony actually turns out, Oteri, Weinberg and Pinales do not investigate the impact of their recommended strategy. One would, however, anticipate that its success would be aided by the unfamiliarity of cross-examination to the scientific witnesses and by the limited distributions of opportunities to speak so that examiners can insist on answers as well as cutting replies short (on these features see Jasanoff 1998 and Chapter 10 of Yearley 2004).

The material to be examined in this chapter allows a sociological development of these themes. The material derives from a hearing about the proposed development of an area of bog for peat extraction for horticultural use. The proposals were opposed by various conservation interests as well as anglers, a guest-house owner and farmers concerned about the drainage implications for their adjoining lands. The details of the examination of witnesses provides a valuable case study in the workings of the kinds of principles first introduced to a sociological audience through the publicising of the work of Oteri, Weinberg and Pinales.

Environment and wildlife conservation in Northern Ireland

Nature conservation in Northern Ireland would, by most criteria, be judged less advanced than in the rest of the UK. A number of reasons can readily be adduced for this. There was first the political turmoil which was at its height during the period when environmentalism was consolidating elsewhere and the related difficulty in introducing conservation legislation when the path of government tended to follow some way behind moves in Great Britain (GB) law. The UK parliament has to enact legislation for Northern Ireland separately from that governing England and Wales, and there are often minor legal differences which hold up the drafting of even essentially identical laws. A further factor was the low priority accorded by government to any issues which were not related to security matters or addressed to problems of employment and the economy in the years preceding the outbreak of the uneasy 'peace' in the mid-1990s. Finally – and ironically – the rural nature of the low-density population in Northern Ireland tended to promote the continuation of the belief that the environment was not much threatened.

One indication of this low priority is the relative paucity of Nature Reserves and other protected wildlife sites in Northern Ireland. In GB the commonest nature-protection designation is, the Site of Special Scientific Interest (SSSI); the Northern Irish equivalent is the Area of Special Scientific Interest (ASSI). In 1985, in the run-up to the inquiry that features in the chapter, the Nature Reserves in Northern Ireland were, size for size, only around 40 per cent the extent in GB and ASSIs had only recently begun to be declared (see Newbould 1987). The apparent Northern Irish lag is partly attributable to the fact that legislation enabling the declaration of ASSIs was not drafted until 1985 and the survey work on which declarations would be based was not begun until 1986 when the staff of the relevant section of the Department of Environment in Northern Ireland (DoENI) began to be increased. The proposed expansion of staff for the surveying work had still not been completed at the time of the inquiry.

Aside from the cases of one or two outstanding natural phenomena, including the Giant's Causeway and Strangford Lough,[2] the basis for the DoENI's conservation work is faced with a difficulty. Briefly expressed, should the aim be to conserve for Northern Ireland, for Ireland as a whole, or in the UK context?[3] Should nature reserves concentrate on what is characteristic or should they aim to include areas which might be better represented elsewhere for the sake of having a Northern Irish example? This is a practical issue in terms of detailing scientific staff to examine different kinds of habitat.

Among the areas that were prioritised early on for potential designation and, in a few cases, actual designation were bogs. They are, so to speak, doubly deserving of conservation since they are clearly a characteristic of Ireland and they have been extensively reduced by working, north and south of the border, both by traditional cutting methods and also using more recent extractive technologies. Irish bogs are held to occur in two principal forms: blanket bogs and raised bogs. In both cases, beds of peat are formed because the watery environment inhibits the usual aerobic breakdown of organic remains. Contrary to what one might expect from their names, it is blanket rather than raised bogs which are characteristic of upland areas. Blanket bogs cover areas where the rainfall is extremely high and/or the drainage very poor. As these bogs develop they tend to drain increasingly slowly. Plant life in boggy areas thus faces a constantly wet environment and must confront the problem that the water from which it gathers nourishment is almost wholly derived from rainwater and is, consequently, very low in nutrients. Moreover, the partial decomposition of the organic material generally produces an acid environment. Being both poor in nutrients and acidic, blanket bogs are unwelcoming for the majority of plant species.

Raised bogs generally form in shallow depressions left after glacial activity. Pools with little or no drainage form in these hollows; they are fed entirely by rain and lose water by slow drainage and evapotranspiration. Being exclusively rain fed and of a limited area these zones may become even more nutrient poor and acidic than blanket bogs. Only specially adapted plants and associated fauna can thrive. Perhaps the most strikingly adapted plants are the insectivorous sundews which augment the meagre nutrient supply to their roots with meaty snacks provided when insects become incautiously tangled in the sticky tentacles on their spoon-shaped leaves.

The 'raised' aspect of these bogs refers to the tendency of the bog plants to grow on top of their predecessors rather like the formation of a coral community. What started as a depression may thus be turned, after thousands of years, into a domed area rising above the neighbouring plain. A typical structure arises with a central area known as the dome; a perimeter or lagg that is often characterised by a stream circumscribing the whole bog and carrying off the water which seeps out; and a sloping area – the shoulder of the dome – known as the rand. Such bogs, being raised above the surrounding land and tending to occur on lowland areas, have been used for turf-cutting for many generations.

In addition to the demand for private, small-scale peat-cutting, the raised bogs are now generating interest among companies that aim to harvest the peat mechanically. Although the traditional use for peat is mostly as a fuel – and this is still the prime use of blanket bog material in places such as the Irish Republic and much of northern Scandinavia – the peat from raised bogs is held to be very suitable for horticultural and gardening uses. Companies wishing to use the resource in this way aim to purchase large tracts of bog. Drainage channels are then dug and the bog is left to drain for two or three years. At this point the top layer of the bog can be broken up with a harrow, the crumbly surface left to dry, and the material 'vacuumed' up by large, tractor-mounted suction devices. At the same time as the peat is being harvested from the surface the underlying strata are continuing to dry. The peat can thus be extracted progressively.

The ability to harvest peat in this way is a relatively recent development (having occurred within the last three decades or so) and was introduced into Northern Ireland in the late 1970s, at least in part with the encouragement of the Northern Ireland administration (see Cruickshank 1987, 106–12; this was stated also by the counsel for the developer, see below). Hence both the peat-harvesting companies and conservation groups had an interest of relatively recent standing in raised bogs. These interests came into open conflict over one particular bog in the north-west of the province,[4] known as Ballynahone More. The peat-working company applied to the Planning Department of DoENI for permission to work the bog, for road access and so on with a view to the production of horticultural products. Objections were forthcoming from many conservation interests and even from another wing of the DoENI itself, the Countryside and Wildlife Branch (CWB), which has a statutory duty to pursue the interests of nature conservation. Accordingly it was decided that the issue should go to a Public Inquiry which was held in December 1987. This case study deals primarily with the inquiry itself though the overall outcome of this episode offers an ironic commentary on conservationists' approach to the inquiry and will therefore be considered in the closing discussion. In the main, however, this chapter focuses not on the outcome of the inquiry but on the nature of the arguments that went on during it.

The inquiry

Public Inquiries in the UK are organised in an adversarial manner and generally resemble legal trials. But as Wynne points out:

> The idea of natural justice and public redress implicit in the litigation model is, however, contradicted by the fact that public inquiries are advisory mechanisms not judicial proceedings, and by their formal role in examining *objections*, with an underlying presumption in favour of development. (1982, 53)

People or companies are free to propose all manner of developments (providing they are not specifically illegal) and then the planning system examines possible objections to them; there is no presumption that the inquiry investigates what the best form of development might be. Only the wrongs and rights of the proposal actually put forward are assessed. In this case, perhaps because of the 'presumption in favour of development', objectors from a conservationist point of view were well represented at the inquiry. In addition to representatives from the CWB of DoENI, there was a senior figure from academic ecology in Northern Ireland, an objection from the local Wildlife Trust and a figure, also a senior academic, from a standing advisory committee to the CWB. The Wildlife Trust called two witnesses (plus a subsequent adviser) and the CWB had a spokesman plus a scientist working on research and fieldwork. These people were generally well known to each other and shared membership of conservation organisations. The developer was represented by the company's MD, a scientific adviser, their drainage engineer, and a Queen's Counsel (QC) or senior barrister and his assistant.

The developer's representative had to face a number of objectors. His position could readily be presented as that of a beleaguered exponent of economic development and job creation. Since so many of the objectors came from government departments, he could also contrast the efficiency of the private sector with the stereotypical lumbering performance of the public services. Apart from the QC, only a very small number of other participants appeared to have experience of adversarial hearings; none of these were from the conservationist bloc. The conservation scientists were overwhelmingly drawn from an academic background; either they were expert by virtue of their academic position or they had recently moved from an academic setting to more campaigning forms of employment. In the relatively restricted social circles of Northern Irish nature conservation they seemed to be accustomed to fairly consensual working and to receiving respect for their work from each other. They seemed quite unused to the prospect of cross-examination and those whom I questioned viewed the proceedings with some trepidation.[5]

The inquiry lasted for two and a half days. Conservation issues occupied something over a third of this time. In broad terms, the hearing was organised so that it started with the DoENI statement of the considerations leading up to the inquiry. This neutral review was followed by the detailed statement by the CWB opposing the development of the bog. The QC then cross-examined all the DoENI participants. This took the proceedings into the second day which began with the remaining questions to CWB representatives. Others could then question the DoENI. This was followed by the company's testimony and the examination of that evidence. Finally, angling and non-governmental conservation interests were permitted to make their presentations on the third day.

The issue of expertise in wildlife assessment arose in three main contexts: the presentations by scientists opposing development, the evidence of the developer's scientific adviser and the cross-questioning by the QC. It is particularly the first and the last of these sources with which I shall deal since, on account of their adversarial arrangement, they serve to display the issues at stake most clearly.

Scientific expertise in the inquiry

Most scientific expertise was arrayed against the developers. Their QC was accordingly obliged to confront the claims of scientific experts to the effect that this bog was particularly valuable and in need of conservation. Some of the basic strategies used coincide with those noted by Oteri, Weinberg and Pinales and can be exemplified in the treatment of the first scientific witness: a bog ecologist from continental Europe who was called by the Wildlife Trust. His testimony came out of turn as he was only able to be present for a short while because of professional commitments at home and arrangements were made to accommodate his timetable.

This witness began by placing Irish bogs in a European context: much of Europe's bog had disappeared and where attempts were currently being made to recover peatlands they were proving vastly expensive. The Dutch had accordingly become involved in fund-raising to purchase Irish bogs for conservation purposes, thus indicating that they were of recognised international significance. He listed six virtues of Irish bogs:

1. Bogs serve as markers of the environmental health of a region;
2. They function as a gene pool for raising strains which are, for example, resistant to acidity or sustained wetness;
3. As peat beds contain a record of past life through the preservation of pollen and other deposits, they are of great value to climatologists, archaeologists and cultural historians;
4. Peatlands are important for education and study, especially as they represent a simple (i.e. with relatively few species) but coherent ecological system;
5. They are an integral part of the Irish landscape and its psychological environment;
6. Boglands and the specific life they support have an inherent value.

Questions to this witness were focused on two issues. First came the practical implications of the arguments he presented. If bogs are as valuable as he asserts, then why not propose a ban on all exploitation? The reply was that we only need representative ones and that since his fellow countrymen had already profited by using their bogs he was hardly in a position to deny this advantage to the Irish. Thus, good though his arguments might be for preserving some peatland they did not have clear implications for the case in hand; some bogs could justly be exploited even in the eyes of this witness. The second issue then followed directly: what are the criteria for deciding which ones to save? What makes this one special?

The initial answer was that the intactness of the bog was the leading consideration. But is not rarity also important? It may be. He was then asked if what he did in his work on bogs was assess their value in relation to habitat preservation. It was not. He did basic research in ecology on 'almost undisturbed' bogs. So he was not an expert in the assessment of bogs nor actually involved in bog assessment in the course of his work? No. And he acknowledged that although he had visited the bog which was the subject of the inquiry he had not visited other raised bogs? Yes. How then, not having specific expertise in bog assessment nor having visited other candidate bogs in the Province, was he able to advise on whether the development of this particular bog should go ahead? No answer that could satisfy the QC was forthcoming.

In this case the questioning functioned along two parallel paths. First, it was suggested that the person was not really an expert on the assessment of bogs even though he might be an expert on bog ecology (although whether there are such creatures as bog-assessment specialists was not determined). Second, attention was directed to the extent to which his views were mere opinion – partly through lack of expertise and partly due to lack of familiarity with raised bogs in Northern Ireland. In other cases also, notably with the people from the CWB, questioning began with the issue of expertise in the field of bog evaluation. The approach is strategically identical to the procedure advocated by Oteri, Weinberg and Pinales: that the witness should be asked for qualifications or expertise in the field of illegal drug identification, a subject which is (of course!) not customarily taught in college courses. The defence counsel does not have to establish that there is anyone who is in this strict sense an expert on drug identification or that unambiguous identification is possible; the task is to show that this witness is not an expert and that his or her identification is not beyond all reasonable doubt.

Wildlife evaluation and scientific conventions

The position of the witnesses from the DoENI's CWB was somewhat different. Because, as noted above, Northern Irish legislation tends to follow some way behind that in GB, the ASSI designation had only been available since 1985.[6] The sole grounds for designating an ASSI are scientific;[7] it may not, for example, be declared on amenity grounds or for other diffusely conservationist reasons. When, therefore, the CWB began to carry out its peatland survey it had to devise a way of operationalising scientific merit. As their submission states:[8]

> The main purpose of the lowland raised bog survey was to identify bogs for declaration as Areas of Special Scientific Interest. An additional aim was to describe and locate the main types of vegetation and physical features on each bog as an aid towards planning the future and conservation of the selected bogs.

The bog whose future was under discussion at the Inquiry was identified for declaration as an ASSI.

At this point the developer's QC might have been expected to develop his argument in either of two ways. He might have suggested that the designation was scientifically justified but that the acknowledged scientific importance was overridden by commercial and employment considerations. In an economically depressed area this might be thought to be a smart move. Alternatively, he might have focused attention on the legal procedures that had been followed in declaring this particular ASSI and sought to find some technical irregularities. However, he chose in the main to look away from economic or narrowly legal issues and to emphasise the details of the designation. The other issues did receive some attention. For example, the QC asked why the designation, once decided, had not been implemented sooner; the answer was that the ASSIs were done gradually to allow time for the necessary pre-notification visits and to spread costs. But the main thrust of the questioning was about the justifiability of the designation itself.

The basis adopted for the task of designating bogs was examined in considerable detail. An appreciation of the issues can best be acquired by examining the nature of the scoring system used in the designation exercise. This is set out in Table 5.1; it follows the format of the CWB's own document.[9]

Using this system of scoring, the CWB had ranked all the raised bogs surveyed. In the 'top ten' bogs, based on scores from these structural features plus scores from a survey of indicator floral species, Ballynahone More came tied-third with a score of 59. The chart topper, with 66, was a raised bog of similar provenance. Another, very similar bog scored 45. Of the top three, Ballynahone More had the largest area.

The document containing these scores was only provided at the Inquiry itself although, since the information was in no sense confidential, it is not impossible that the developers could have known of it beforehand. In any case their QC was quickly able to build up a critique

Table 5.1 A 'scoring system' for raised bogs

Structural feature	Score							
	0	2	4	5	6	8	10	15
Intact surface as % of original bog	0–25	25–50	50–75	–	>75	–	–	–
Intact lagg as % of original	0	–	–	1–33	–	–	33–67	67–100
Sphagnum (bog moss) cover on intact surface (%)	10	11–20	21–40	–	–	>40	–	–
Internal drainage system	Absent	–	–	Present	–	–	–	–
Development of hummocks & hollows	Flat	Poor	Fair	–	Good	–	–	–
Development of pool system	Absent	Poor	–	–	Fair	–	–	–
Area of intact surface	1 point for every 10 ha							

of the table and the procedures underlying it. For analytic purposes four strands can be identified in that criticism although they were not used independently.

The first approach depended on the identification of inaccuracies. Thus the company too had come up with a measure of the extent of the bog using similar aerial photographs and similar equipment for reading off areas. However, this estimate set the size of the intact bog at 95 rather than 105 hectares. The developer's representative suggested that, given the company's financial interests as opposed to the less acute concerns of the DoENI, they would be more likely to have got this right. (A later witness testifying for the Wildlife Trust set the figure at around 90 hectares.) This error, if such it was, not only served to throw doubt on the Branch's technical competence in general but also would reduce the score of the bog by one point. In a similar way the reading of the pool complex (rated as 'fair') was challenged on the grounds that when the bog was surveyed it was a particularly wet month. Again the company should know since the weather has direct commercial consequences: that month it was so wet that harvesting operations were halted elsewhere. Could the assessment of the pool and other phenomena not depend on such contingent matters also? In a month with average rainfall, maybe the pool system was more like 'poor', worth four points fewer.

The second kind of doubt was thrown on the scores in relation to marginal cases. If one looks, for example, at the first structural feature (the intact surface as a percentage of the original bog surface) it will be seen that points increase in steps of two between rather broad percentage boundaries. Surviving areas of both 26 per cent and 49 per cent will earn the bog 2 points, while a surviving area of 55 per cent would gain it twice as many points. It so happens that for the bog being considered the estimated intact surface is 51 per cent of the original. But, it can be asked, does that single percentage point justify the extra two added to the score? Of course, this kind of point can be run in parallel with the first to indicate that, when a better estimate of current size is obtained, the intact surface will be shown to be a smaller proportion of the original bog.

A third argument functioned by casting doubt on the independence of the features enumerated in the table. Did not the procedure of scoring both the absolute size of the intact bog and its proportion of the original surface overvalue size as a feature? Or, the other way round, did not the lagg receive a poor deal from the scoring method? In 'reality' the lagg is of a twofold importance; both for the characteristic flora which exists only in the specific environment of the bog's edge and for the lagg's own role in sustaining the hydrological integrity of the bog. Yet it was counted only once.

Although the third strategy was of relatively limited application the fourth was of great importance; it depended on questioning the scores attached to the different features. As one can well imagine, it is easy to question why a 'fair' pool system has one and a half times the advantage over a 'poor' one that a 'fair' hummock development has over a 'poor' hummock complex. Equally, why should a score of eight points be awarded for one thing only: a sphagnum (bog moss) coverage exceeding 40 per cent? The QC noted that, in his view, the CWB representatives behaved as though the scores were 'written on tablets of stone'. But if they are not prescribed in some similarly unquestionable way, one can ask whether, given a different set of weightings, the bog would not in fact be shown to be rather a poor example. The case for doubting the outcome of the scores was ironically strengthened in subsequent testimony when one of the academic ecologists argued against exploitation of the bog, stating that his opinion was that this specimen was a more valuable bog than the top-scoring ones. That testimony may have seemed a good idea at the time but it did little to support the value of the Branch's 'systematic' assessment work. Although he did not say so expressly, it appeared that this ecologist placed great importance on

Ballynahone More because of its large size. Or to put it another way, he rated size above the structural features like the intact lagg which accounted for the other bogs' high scores.

In its submission the CWB had actually acknowledged this difficulty. For each bog the scores were totalled: 'to give an index of scientific value. It must be stressed that an index compiled in this way has no absolute meaning; it is no more than a rather crude device to bring together a set of observations'.[10] Although this statement indicates that the problem was not unanticipated, it does not provide an answer other than that the score was inevitably in some sense a convention or artefact. Of course, this 'admission' is extremely familiar from the sociology and philosophy of science, where the 'conventional' nature of scientific agreements has long been noted (see Jasanoff 1996 for example). But the legal examiner was able to use this quote to suggest that there was no coherent basis for the scoring which, to all intents and purposes, could be re-weighted to favour the choice of any bog. He also made a point of asking whether the criteria were established in international agreements or some similar precedent. However, since the criteria had been specially adapted for their current use, this route to legitimising them was undermined also.

It should be appreciated that these objections could have been raised to any scoring system. Admittedly, the one in use probably did live up to its description as 'crude'. For example, the document in which it is described makes no mention of tests of robustness. One might, for instance, have expected there to have been trials to see what the effect on the rank order would be of small changes in the system of scoring. If such checks were run, there was no mention of it in the document or in the testimonies. But, unless the top bog had come top on every single conceivable measure, the second bog second on every count and so on (and this, of course, excludes the element of interpretability in attaching scores to real-world phenomena), any scoring system could have been similarly deconstructed. Consequently the CWB's situation was rather that of being 'damned if you do and damned if you don't'. Had they not tried to devise an objective instrument and just based their assertions on their 'impressions' or 'judgement' they could readily have been attacked for being unscientific and subjective.[11] One fieldworker's assessments might then have been held to differ radically from his or her colleagues' views or even to vary from day to day. On the other hand, any instrument – however surrounded by caveats – is bound to commit them to some system of scoring which lacks transcendental validity and which can be made to appear arbitrary if not tendentious.

The QC's critique was assured of (a measure of) success. What it demanded for this success was both a close appreciation of the use of 'scientific' methods of assessment and the invocation of an inflated image of scientific objectivity.

Ad hoc adjustments in wildlife evaluation

A further major difficulty to be experienced by scientific witnesses during cross-examination was also foreshadowed in the CWB's submission. Immediately after the passage cited above it was asserted that 'the numerical index value obtained for each site must be interpreted with regard to the key features of each site'. In other words, to use the index in a practical way, it has to be judged in the light of expert opinion about each bog.

In practice, the significance of this claim could be developed in different ways. For instance, government conservationists on a site visit might detect some feature that is of great importance but which is not sufficiently general to be included in the index. In this case such an argument was mounted on behalf of the Large Heath butterfly. A written testimony from a zoologist was brought forward stating that the bog was particularly important as a breeding site for this butterfly. Although found over a wide area of Europe, the butterfly was apparently in evidence on a diminishing number of sites. Accordingly, the fact that it happened to breed on this bog should be taken as an additional factor, making the bog even more important than its score indicated.

Yet once this possibility is admitted in, so to speak, a positive sense – when an unexpected but important faunal item is found – it can also have a negative side; for just as a medium-scoring bog may be raised to importance by the presence of a special feature, so also may a high score be discounted if other aspects of the bog are inauspicious. Thus it was put to the CWB's witness that the intact surface of the bog, although large, was 'linear' (i.e. not circular or even approximately so). The witness accepted that it was 'somewhat elongated'. It was then argued that since the lagg was all cut away (Ballynahone More scored zero on this feature) the linear shape dictated that no part of the dome was very far from a cut edge and was thus likely to drain and dry out even without the developers working it. It was a 'doomed dome'.

Although this argument was damaging in itself (in that it held out a gloomy prospect for the bog's future), it had an additional significance. It was part of an attack on the way in which the apparently 'scientific' method of scoring had to be modified, even manipulated, in the light of *ad hoc* considerations – considerations which did not feature in the assessment system itself.

Other features, exhibiting a second kind of 'ad hocery', were drawn on in building this attack. Thus, it emerged in the course of cross-questioning that one bog with an intact lagg had suffered some digging on the lagg itself although it appeared still to be functionally intact. Did this really count as still intact? When the reply came that the lagg was rightly to be regarded as intact even though – literally speaking – it had been dug, this led to the question: 'wasn't that very unscientific?' The criteria, it was said in reply, have to be interpreted. But, of course, if the criteria have to be interpreted, this act of interpretation can be presented as very problematic. If it all depends on interpretation, the scoring system hardly appears to act as a constraint at all. It seems to be a device which conservationists can invoke when it suits them and which can be overridden when it does not suit.

The vital importance of expert interpretation rose to prominence over the issue of previous manual cutting on raised bogs. Cutting done in this way tends to produce a worked face along part of the bog's perimeter rather than a progressive cutting downwards in the manner dictated by the new machinery. Cutting thus has the effect of destroying at least part of the lagg and produces a surface from which water can drain out of the bog's internal structure. Peat-working at Ballynahone More over the years had been extensive although the sheer size of the original bog meant that a good deal of dome remained. But estimations of the future of the bog as a hydrological system depended on whether and how quickly the bog would dry out because of the previous cutting. The CWB's witness suggested that the cutting had not been very deep; only shallow faces had been cut so far. This would indicate that there would be less drying out than would occur on many other traditionally cut bogs. But despite repeated promptings the witness would not put an estimated value on this effect. He would not, for example, offer a value representing the effect of a five-foot (1.5 m) face on the water 100 feet (30 m) into the bog. Because he would not attempt to put a value on the drying-out effect of traditional cutting (the cutting which the developers insisted doomed the bog to drying out) but would only say that the relatively shallow cutting was very unlikely to be terminally harmful to the bog, the developer's QC accused him of taking shelter behind 'scientific privilege'. He wished – it was implied – to use his scientific standing to say that the effect was negligible but would not, or could not, give any useful indication of the size of the effect.

This whole line of argument about the role of *ad hoc* and unquantifiable assessments was admittedly 'licensed' in some sense by the attempt to boost the bog's value by invoking the Large Heath butterfly. But it appears to function by running together at least two sorts of *ad hoc* reasoning. The first corresponds to the addition of new criteria such as 'butterfly-supportingness'. The second relates to the role of scientific judgement in bringing untidy empirical observations into accord with any set of formal criteria. In the sociology of science literature the importance of scientists' tacit skills and of the negotiations which link the messy world of observing to formal records of those observations has been repeatedly stressed (see Collins 1992). Yet here the QC treats these ineliminable features as though they were deviations from scientific procedure. The informal skills which make science possible are treated as though they were, possibly tendentious, short cuts. Once again, the apparent coherence of the QC's attack seems to trade on an inflated image of scientific objectivity.

Contesting conservation value

My initial analysis of this case was undertaken before the outcome of the inquiry was officially announced. It appeared to me from my observation of the inquiry process that the QC's strategy was a winning one. He managed to keep the issues of scoring and evaluation as the central themes. Within this pivotal area he was able to throw doubt on the actual calculation, the assumptions which lay behind it and the expertise and care which had gone into its use. Moreover, a detailed examination of the use of the scoring system in practice allowed him to focus attention on the supplementary and *ad hoc* reasoning that permitted the system to be applied to a diverse and unpredictable world. While such ad hocery was no doubt reasonable, it could not by its very nature be built into the scoring system. It could thus be presented as the arbitrary use of scientific privilege. Equally, it provided an opportunity for the developer's representatives to bring in new elements themselves, such as the linearity of the bog, which – it was then implied – had been carelessly or tendentiously omitted by the conservationists. My expectation was wholly compatible with the outcome of the process. The Planning Commissioner (the executive judge in this inquiry) recommended that planning permission be granted; his advice was accepted by the relevant government minister and reluctantly accepted by the CWB.

The QC's double task was to render the scientific evidence pertaining to the case intelligible so that the Commissioner could pass judgement whilst also attacking the specific scientific expertise of his adversaries. In working towards this public understanding he was at pains to distance himself from scientific learning; for example he claimed to have acquired all he knew about raised bogs from the publications of the DoENI and of conservation groups, and he rather theatrically looked to others for the calculation of even simple sums like the percentage of the original bog surface which still remained intact. At the same time, he invoked an everyday understanding of the nature of science in general when, for example, he accused witnesses of having acted unscientifically or of hiding behind scientific privilege.

A number of features of the legal context assisted him with this task. First, as has already been noted, the objectors were appealing against the development; he was not having to plead for it. Consequently, the developer's QC had only to find weaknesses in the evidence for the case against exploiting the bog. He did not, for instance, have to rank the bogs or show that an agreed ranking was conceivable; he only had to undermine the objectors' ranking. His argumentative work was thus conducted in a thoroughly asymmetrical context.

Second, the argument was being conducted for an audience. Unlike a professional scientific debate or even many open policy disputes, the objective was not to win opponents over, to persuade their supporters away, or even to enrol new allies. Instead the aim was to discredit the opponent's presentation in the eyes of a third party. This, together with the legal rules distributing opportunities to pose questions, lent the debate its uncompromising character. This feature was compounded by a third factor, the limitation on time. The planning decision was going to be made on the basis of the few days' proceedings. In routine scientific controversies scientists can often shelve disagreements and anomalies, sometimes for years or even decades. But in this case there was no opportunity to refine the procedures or to re-evaluate disputed results. Thus, while many of the challenges posed by the QC are structurally equivalent to argumentational strategies adopted in scientific controversies, the opportunity to revise one's arguments and the chance to invoke counter-challenges were both absent (see Lynch 2002 for parallels). While, as Wynne has observed (1982, 129–37), science is often treated by legal practitioners as – in principle – the supplier of uncontested truths, the circumstances and conventions of legal debate appear to discourage the sensitive application of scientific expertise to whatever matter is currently in hand.

Overall, it is clear that the features which the QC highlighted to his greatest advantage – the conventionality of the scoring system and the dependence of scientific categorisation on tacit skills and unexplicated interpretation – are the very aspects which authors from the sociology of science have insisted typify all scientific knowledge (see Yearley 2004). But in this case the QC was able to mobilise a harsher, more scientistic standard against which the scientists' practice could be unfavourably compared. In this case it appears that the conservation scientists fell prey to an exaggerated image and expectation of science, an image which scientists have themselves often used to impress on the public the importance and authority of science (Yearley 1994).

However, the story finishes with one final irony. In the 1990s, Friends of the Earth (FoE) began operating in Northern Ireland with greater effectiveness following the setting up of many local groups and the eventual appointment of a full-time campaigner and development worker. This worker re-examined the case and, in consultation with others, elaborated the following argument. The DoENI, once it was satisfied that Ballynahone More was of ASSI standard, had been obliged to declare it an ASSI and to protect it, even if that meant that it had to pay compensation to the licensed peat-harvester. Had the inquiry gone the Department's way, they would have had the bog preserved costlessly. But even now – so FoE argued – they had a legal duty to conserve it; it was after all a high-ranking bog on their own scoring system.[12] FoE (England, Wales and Northern Ireland; Scotland has its own FoE) threatened legal action and the issue was quickly resolved. In effect, the company traded the bog – which it owned – for comparable peatland of less wildlife value, and the DoENI secured Ballynahone More for less than it would have had to pay in compensation. The true irony in all this of course is that what the conservation scientists lost in their legal battle in the inquiry was virtually recouped by environmentalists' own legal action. The successful strategy paid little attention to science and a lot of attention to law. Compared to US environmental organisations, those in the UK are scientifically highly qualified and legally under-qualified. Curiously, a US-style approach might have served them better in this case.

6
Independence and Impartiality in Legal Defences of the Environment

Introduction: expert advice and the funding of environmental movement organisations

Voluntary organisations concerned with environmental protection and nature conservation clearly need a source of funding to pay staff, to undertake their projects and for various administrative tasks. Some depend almost exclusively on donations and membership subscriptions and devote a great deal of organisational resource to funding drives. They may also trade in books and tee-shirts, even seek commercial sponsorship. Others receive governmental support in various ways. For example, the Royal Society for the Protection of Birds (RSPB) receives official grant aid towards specific projects, funding which complements its highly successful private and commercial fund-raising operations.[1] Exploiting a different opportunity, the various Wildlife Trusts that made up the Royal Society for Nature Conservation (RSNC) used to get a good deal of Manpower Services Commission money to support workers under the 'Community Programme' job-creation scheme which ran in Great Britain during much of the period of highest unemployment in the 1980s. And since the government's main aim in distributing this money was to get recorded unemployment figures down, the Trusts felt reasonably free to allocate the money according to their own criteria. Moreover, the groups in the RSNC have traditionally been close to the government's statutory nature conservation body (at the time of the case study, the Nature Conservancy Council, NCC[2]) and many of them do the kind of work which the NCC or – in Northern Ireland – the DoE Countryside and Wildlife Branch (CWB) 'should' be doing.[3] Accordingly, for a period in the Thatcherite 1980s and through into the early 1990s, several Trusts had little difficulty in obtaining basic funding, especially since they tended to do the conservation management work more economically than the official bodies could do it themselves.

However, a high degree of reliance on such official largesse is felt to compromise these groups' independence. They too thus look to other fund-raising ventures, such as sponsored tree-planting and the sale of various goods, from calendars, Christmas cards and books to more exclusive items such as wildlife holidays. Given the high standards of scientific expertise commonly available in these conservation organisations, they are increasingly attracted to selling their skills in the form of consultancy services – to firms, to developers and to local authorities.

This commercialisation of skills may take a number of forms. They can gain money from selling the kind of advice that they might anyway have found themselves giving: to farmers or park-keepers on habitat management, to industrialists on landscaping and greening of their grounds, and to education authorities on the greening of school premises. They can tender for environmental landscaping contracts, which might otherwise go to commercial concerns, many of whom have now cultivated their 'environment-friendly' profiles. They can even sell 'eco-audits' of a company's or a local authority's plants, examining the energy efficiency of their lighting and heating, checking their recycling policy and so on. The company gets to be 'greener' and receives a certification of its 'greenness' while the charity gets additional funds. But with EU legislation now requiring Environmental Impact Assessments before large developments go ahead, there is a role for consultancy services in advance of actual developments, to establish – for example – the best location for a hotel or the best design of a jetty and to identify and then minimise the likely ensuing environmental problems.[4] It is this area – both lucrative and environmentally important – which is attractive to environmental consultants, including those developed from Wildlife Trusts. In particular, the Trusts are keen to do this kind of work because they believe that they have nature's interests at heart in a way that commercial concerns possibly do not. As one respondent expressed it: 'We do it not only for the fee but to make sure that good work is done.' It is an example of work of this sort, an impact assessment prior to development, which forms the case-study material for this chapter; for while this strategy is already yielding some financial successes, it generates a potential problem: how can Trusts demonstrate that the advice they give is truly impartial?

The case study: competing marinas

Northern Ireland is world famous, not for its wildlife but for its political problems and also for its economic malaise arising from the collapse of shipbuilding and textile industries. These things are not unrelated. Want of industry and development has been responsible for the comparative lack of pollution in the Province.[5] At the same time the urgency of planning control has not been felt and unattractive developments have gone ahead, particularly in rural areas. The NCC equivalent, the CWB – part of the Department of Environment in Northern Ireland (DoENI) – has been nothing like as pervasive in its influence on nature conservation as the NCC in GB. As part of the civil service, the CWB is ultimately subordinate to the larger objectives of government policy, namely the maintenance of security and the promotion of economic development. And, since the government is also the biggest employer in Northern Ireland, its impact on the economy and thus the environment is vast (see Yearley and Milton 1990).

At the same time, there is a great deal of unspoiled countryside in Northern Ireland, and outdoor leisure pursuits are popular, not too exclusive and readily available. Sailing is a particular favourite and, because of low house prices during 'the Troubles', those in work were probably more able to afford a yacht than would be the case elsewhere in the UK. There are a number of yacht clubs and a few marinas.

One particular site favoured by yachting enthusiasts is Strangford Lough, a sheltered marine loch (lough) off the Irish Sea, south-east of Belfast. The lough is large, well sheltered and comparatively shallow. It is bordered by few towns and has a rich and diverse marine life. Much of the coastline and some islands are owned by the National Trust (on the National Trust, see Pye-Smith and Rose 1984) and the entire shoreline, up to high-water mark medium tide, has been declared an Area of Special Scientific Interest (ASSI) (the Northern Irish version of the SSSI, Site of Special Scientific Interest, the principal wildlife designation in GB). The shelter and scenic attractions which Strangford Lough offers mean that there are a number of would-be users. People wish to fish it, to go diving in it, to farm fish in it, to sail steamers on it and so on, and conservation and amenity groups have grown used to having to defend it.

The case study to be considered here concerns plans to develop yachting facilities in a small town (with around 2000 inhabitants) – Killyleagh – which lies on the shores of the lough. The town has its own yacht club and boats are moored along the stretch of coast immediately to the south of the town. Several other yacht clubs exist around the lough but there was at this time no marina; recently developers had been keen to press ahead with this possibility, arguing that it would combine economic development for the area with a tidying up of yachting in the lough. By summer 1989 there were three rival schemes before the planning authorities. Now, the DoENI had already published a plan outlining the general requirements for development of the lough and some of these constraints must be quickly outlined (see DoENI 1984). Development around Killyleagh was limited by three designations: the ASSI on the lough shores (already mentioned), a proposed development limit around the town itself, and the declaration of most of the town as well as the countryside around it as an Area of Outstanding Natural Beauty (AONB).[6]

Turning to the proposed developments, the three plans were as follows:

A. A plan for a marina and hotel complex just north of the town on a green-field site;
B. A plan for a small marina in the existing harbour together with an extension of the existing (small) quay to permit the use of larger cargo vessels, for the export of stone from local quarries;
C. A plan for a marina with a club-house and some associated housing and holiday accommodation to the south of the town.

Each scheme required development on an ASSI and within an AONB. Additionally, the first and third proposals took development beyond the planning limit for Killyleagh while the second would introduce large, commercial boats to the harbour. The first plan involved development which would lie wholly outside the development limit while the third included residential accommodation, a development specifically discouraged in the North Down Area Plan (see DoENI 1984, section 41.3.i, page 69).

The three plans were taken together to a public inquiry held in Downpatrick, the county town of County Down, in 1989.[7] At this inquiry, in addition to various representatives from the DoENI (including the CWB), there were representatives from the three companies and their barristers, objectors from three environmental organisations and a local solicitor representing townspeople's interests – largely concerning anticipated noise and traffic problems. The environmental groups were: the National Trust, the RSPB, and the Ulster Wildlife Trust (UWT), the local Trust affiliated to the RSNC.

The developers were all aware of the environmental considerations that were likely to figure. Plan A had been mooted some time in advance of the inquiry and environmentalists had shared anxieties about it. Each developer was at pains to stress that his scheme was environmentally benign, that it disturbed wildlife but little and that it caused no impairment to visual amenity. Realising that environmental issues would assume a high profile, even though the developments were too small to require an Environmental Impact Assessment under NI law, the developers had retained the services of various consultants in advance of the inquiry. The developers of plan C turned to the consultancy arm of the UWT (subsequently termed Ulster Wildlife Consultants [UWC]). It is the role played by the Trust, its consultancy arm and its relationship to its customer that will form the focus of the case study in the rest of this chapter.

The Trust produced two documents: (i) a 'Preliminary environmental impact assessment' carried out for developer C, stated to have been authored by a senior figure in the Trust and formally part of that developer's submission to the inquiry and (ii) its own submission, ascribed to a scientific officer of the Trust.[8] From the wording of these documents and from interviews with the Trust's representatives, it is clear that the Trust's working assumption was that these two documents reflected different and distinct approaches to the inquiry. One was intended as a factual report on the likely impact of the marina development C; the other represented the Trust's opinion on the desirability of each of the developments, including plan C.

At the inquiry the Trust could have been called to speak to its proof of evidence for the developers (in the consultancy mode); they also had the opportunity to question witnesses and to make objections in their role as an environmental campaigning group. They had two representatives at the inquiry and, interestingly, discussed at the outset which person should occupy each role, even though names had been attached to the written submissions. This was most likely because the Trust's 'opinion' submission, the document which might receive the stormier treatment in the inquiry, had been ascribed to the more junior and less experienced employee. In the event, the persons named largely spoke to their own reports. At times – because the proceedings lasted several days and the individuals had other work to do – both officers spoke in the capacity of objector. That the personnel were treated by the Trust as interchangeable indicates that the roles were not seen as personal but functional, corresponding roughly to 'factual talk' and to 'judgement'.

Passing for impartial

The representatives of the UWT/UWC experienced a difficult and unsatisfying time at the inquiry. Initially the prospects had looked good for them. Although much poorer than the National Trust or the RSPB, the UWT alone among the environmental groups present had earned income from the process leading up to the inquiry. Their consultancy wing had achieved a success in being appointed by developer C in the face of commercial competition. They also felt that their representatives were well informed about the site, not least because they had been able to study it a good deal in the course of their consultancy work. But, during the inquiry itself, the Trust suffered several setbacks. Their scientific officer, speaking for the Trust in its pressure-group role, was subjected to very hard questioning by one of the developer's (scheme A's) barrister. The inquiry's commissioner also ruled that the Trust, *qua* pressure group, could not put questions during the presentation of the UWC evidence for developer C. According to subsequent minutes of a Trust meeting, it was generally felt that this levelled an unfair criticism at the Trust, suggesting that the giving of advice through consultancy biased any objection that would be made against developers that hired their services.[9] Coverage of the inquiry in at least one local newspaper was unflattering to the Trust; this was acknowledged in the minutes cited above where the opinion was recorded that 'the public conception of the Trust's position was clearly damaging'.

Linking all of these setbacks was the issue of the Trust's apparent or perceived impartiality. The commissioner's ruling on cross-questioning was felt to impugn their objectivity, while developer A's barrister sought to call their disinterestedness into doubt. In the following sections, I shall examine the bases for these challenges to the Trust's impartiality.

Aspects of impartiality

The Trust's objectivity and impartiality were essential to the simultaneous roles – of independent expert and objector – that they were seeking to fulfil at the inquiry. On the one hand, to be successful consultants it was necessary that the report prepared for their client by UWC should contain the information that any expert wildlife consultant would have produced. The contents of the report should not be covertly shaped by the Trust's own conception of a desirable outcome; it should simply document the consequences that would follow from the developer's own plans, for example in terms of additional water pollution or habitat disturbance during the construction phase. The Trust's sensitivity to the needs of wildlife would ideally ensure that these consequences were documented exhaustively. On the other hand, the Trust prepared a submission reflecting its role as a guardian of the natural environment. In this document they put forward arguments about the way that the consequences of the various developments should be evaluated, recommending for example that the economic benefits supposedly documented by other kinds of experts should not be accorded undue priority over environmental values. In other words, the key to their twofold appearance at the inquiry (and indeed the key to their viability as objective consultants) was their ability to separate their scientific assessments from their value-based recommendations.[10] It was this separation which came under persistent attack during the inquiry.

The first problem facing the Trust's attempts to keep questions of fact and of value separate arose in relation to the opinions espoused by the other conservationists present. The three non-governmental environmental bodies had come to the inquiry with different attitudes to the proposals. Thus, while the National Trust objected strongly only to plan A (chiefly on amenity grounds and citing the AONB designation and the development limit), the RSPB opposed all three (stressing the ASSI designation). The RSPB argued that the preservation of the integrity of the lough's environment demanded that any development on an ASSI be opposed. For its part the UWT was strongly opposed to A, because of its green-field location, and to B on account of the large boats which it would encourage into the lough. It more or less shared the National Trust's lenient attitude to development C. However, given that all three were reputable conservation organisations yet they had come to three different conclusions, it was hard to argue that environmental values alone could have led the UWT to oppose A and B while seeming to condone C. This opened the Trust to the suspicion that its judgement – which happened to favour the consultancy arm's commercial client – was tendentious. This doubt raised a problem for the perceived overall integrity of the Trust as a campaigning organisation and for its public standing as a body with an independent outlook, an outlook founded on its respect for nature conservation. One could additionally infer that if even the campaign arm of the Trust was soft on its client, then the UWC's factual evidence was likely to be compromised too. Of course, this difficulty became particularly apparent because – somewhat unusually – the inquiry was dealing with three rival schemes rather than with the assessment of a single planning proposal.

As I have already mentioned, the difficulty of maintaining the integrity of the Trust's evaluative stand was compounded by a procedural ruling of the inquiry's commissioner. During the proceedings each objector was allowed to pose questions to each of the representatives of each of the developers (thus, for example, the RSPB was permitted to question developer A's engineering consultant). Subsequently, each of the developers was permitted to examine the submissions made by each of the objectors (for example, scheme B's barrister was able to question the National Trust's representative). However, the UWT was not allowed to cross-question scheme C. Specifically, the commissioner would not permit the Trust pressure group to question the Trust's staff member who was acting as the representative of UWC. For that matter, he also forbade cross-questioning of the Trust (as an objector) by scheme C's legal representative. The commissioner said that to allow the UWT to question the UWC was to allow the Trust 'two bites at the cherry'.

The UWT's representatives were dismayed by this ruling. In their view, the scientific survey conducted for the company was fully separable from their judgement about the desirability of the development from a wildlife-conservation standpoint. To them, the ruling seemed unfair because it prejudged the question of this separation. The Trust was not even allowed to make a case for the two aspects being separable; they were simply ruled to be inseparable. The Trust's representatives were further alarmed when, during a break in the proceedings, developer C's barrister – in reference to this ruling – joked 'Oh, but I thought we'd bought the UWT'.

Impartiality under examination

The Trust's general attempt to preserve for themselves a privileged position of scientific disinterest came under attack in a different way during cross-examination. Developer A clearly faced most opposition from the conservation lobby; only his scheme was opposed by all three conservation groups. Accordingly, his legal representative dedicated a good deal of effort to addressing the conservationists' objections. His approach to the UWT depended essentially on two kinds of argument. The first of these involved stressing the extent to which the Trust's judgements were supported by considerations other than purely ecological or strictly scientific, natural-historical ones – that is, considerations outside those over which the Trust might be presumed to have expertise and some sort of mandate.

In preparing this argument he began by citing the Trust's own submission in which it was stated:

> The Trust feels that full support is not justified for a plan to develop outwith the pre-designated development boundary for Killyleagh

> stated in the Down Area Plan for Killyleagh [a companion plan to the one cited above]. We do feel however that this particular proposal [scheme C] constitutes a development that is in keeping with the spirit of said plan.

And in summary their general position is stated as being that: 'Guidelines as laid down in previously agreed Area Plans should be adhered to, at least in the spirit in which they were intended' (UWT 1989a, 7, 9). In other words, the Trust had argued that scheme C – while not enjoying their full support – could be endorsed by conservationists since it accorded with the spirit of the relevant development policies. Drawing attention to these comments, A's legal representative cited two 'guidelines' from the relevant Area Plan for Strangford Lough as a whole:

> 41.3 Planning permission for new dwellings will normally be given only to those who need to live in the rural area because of the nature of their employment – permission will not be given for groups of dwellings.

While:

> 41.4 sympathetic consideration may be given to projects designed to cater for outdoor recreational activity or to facilitate the tourist industry. (DoENI 1984, 69)

Surely, he was able to suggest, if planning concepts are going to be invoked at all, the *spirit* of these explicit recommendations would lead one to expect the following. While his scheme (A) was likely to 'facilitate the tourist industry' and could thus expect to receive 'sympathetic consideration', only the refusal of planning permission could be anticipated for a scheme (like C) that included housing for non-locals.

Since the UWT's respondent was inexperienced (see above) and poorly prepared for this line of questioning, the answers supplied were hesitant and unconvincing. The Trust's representative was only able to suggest that these guidelines were not to be taken as strict rules and, consequently, that the Trust's argument about the 'spirit' of the policy still stood. None the less, the barrister made the point that scheme A appeared more in line with the guidelines than did C. This was not only a positive point for his client but also served to suggest that the Trust was poorly acquainted with planning policy and its procedures. From this position he was then able to imply that the Trust's appeals to 'the spirit' of Area Plans plainly took them beyond their claimed specialist competence. Then, from this point – that the Trust was jumping to conclusions not justified by its particular area of expertise and competence – it was a short step to the inference that the Trust was being partial to its client by choosing to see scheme C as being in accord with the 'spirit' of the planning constraints, while the rival plans were not.

This line of questioning thus delivered three simultaneous advantages to this barrister's client. It afforded a favourable comparison between schemes A and C, it diminished the force of the Trust's opposition by making the organisation appear less than professional, and it invited the inference that the Trust was using planning terms as a rhetorical justification for endorsing the scheme of its commercial arm's paying client.

The barrister's second method of attack on the Trust dealt even more directly with the question of impartiality. Examining the Trust's representative about the UWT's objections, he inquired whether the commercial branch of the Trust had discussed their report with the parts of the UWT responsible for lodging the objection. The witness equivocated, saying it had been discussed but that they (the objectors) had not seen the full UWC report. He asked if the staffs responsible for the consultancy and for the objecting were separate. The answer was that they were not fully separate. It was asked whether the potential conflict of interest been discussed? 'Yes it had been' was the answer, but the staff of the Trust had felt able to go ahead, separating out the functions. Finally, the Trust's representative was asked what influence had been exerted on the views of the Trust pressure group by the fact that its consultancy services had been retained by one of the developers. The answer was, 'very little'.

Clearly, the Trust's demonstrable impartiality was threatened by these answers. It appeared that the organisation had not taken the necessary institutional steps to separate its supposed two functions: the two reports (one for the developer, one for the inquiry) had not been kept separate and confidential; the staffs of the UWT and UWC were not discrete (even if they had been discreet). Even the final reply that the commercial interest had exercised but 'very little' influence could be heard as implying that it had at least exercised *some*.

This line of questioning succeeded in throwing doubt on the facts/judgement distinction. It very clearly left the Trust open to the charge that the views it espoused as a campaigning group had been affected by the requirements of its commercial sponsor.[11]

Discussion: the sociology of impartiality

One way to view this inquiry is as a practical playing-out of disputes about the impartiality of scientific advice. A proponent of the 'received view' of natural science would argue that of all the types of cognitive authority available, scientific advice would be the most perfectly impartial (see Mulkay 1979, 1–26). More recent sociological studies of the role of scientific expertise, particularly in public controversy, have highlighted specific weaknesses in the practical authority and decisiveness of scientists' knowledge and testimonies (see Nelkin 1979; Smith and Wynne 1988). An emphasis on the socially constructed nature of scientific knowledge claims makes it less surprising that scientific expertise can be called into doubt. Thus, since evidence is now believed to underdetermine scientists' interpretations, it is no longer surprising that expert views can be contested. Equally, since scientific theories and generalisations are recognised to be revocable, it is no surprise that courts can treat expert witnesses with a great deal of scepticism. All the same – even given the wide, and growing, acceptance of social constructionist interpretations of science (Yearley 2004) – it is worth taking time to examine the various factors which permitted 'impartiality' to be rendered problematic in the course of this inquiry.

First of all, did the Trust's staff do something wrong: did they, for example, simply organise themselves badly on this occasion? Certainly, they were not as well prepared for several lines of questioning as they might have wished to have been. But we should be clear that there is no straightforward set of steps that they should have taken, which would have secured their claims to impartiality. For example, even having a very rigid and formal separation between the UWT and the UWC would not have proved that their policies had not been informally co-ordinated. In the inquiry, the conservationists in the RSPB were permitted to ask questions of the Trust UWC. There was nothing to stop the UWC planting questions with the RSPB; there are clearly close links of friendship and mutual purpose between the two organisations. Such putative links could also have been drawn in during cross-questioning.

The same would have been true had the Trust followed a policy suggested at a later meeting of the UWT's conservation committee: that the consultancy arms of the various Wildlife Trusts (since the UWT was not alone in seeking revenue from this source) operate exclusively *outside* their own county areas. For example, the UWC might consider operating in Scotland while the Scottish Wildlife Trust's consultants could operate in Northern Ireland; Avon could sell its services in Staffordshire and vice versa. On the face of it this might remove some of the more obvious problems facing the witnesses but again it would by no means be foolproof. There is clearly no specifiable set of criteria which would ensure impartiality and freedom from charges of improper familiarity. In an adversarial setting, where it is in one party's interests to imply that the consultants and objectors are partial, no measures could guarantee freedom from accusations of partiality.[12]

A second impediment to demonstrable impartiality derives from pragmatic considerations about the inquiry process. Very many conservationists, including members of the Trust, thought it extremely likely that some marina development or other would go ahead in the lough. Indeed, in both their reports the UWT acknowledged that it might in some respects be an environmentally beneficial development since yachts could be regulated and controlled effectively in a marina (UWT 1989a, 2, 1989b, 4). Their view was that scheme C was probably the least bad from a wildlife point of view. They were then faced with the practical difficulty of squaring this preference, which they had arrived at from a nature conservationist's point of view, with the existing planning regulations, which involved many other kinds of considerations as well (for example, amenity, landscape and so on).

In these circumstances, objectors face a dilemma. They have to present their arguments so as to reflect their constitutional role: as a nature conservation organisation committed to upholding and improving legal safeguards for wildlife. At the same time they need to make predictions about the likely outcome of the inquiry (i.e. that some sort of marina would probably be built in the end) and to use their arguments to try to influence the kind of planning consent that would be given for such a development. In this case, this led the Trust to adopt a rather equivocal attitude to the possibility that a marina would be built. The developers who faced most opposition from environmentalists were able to exploit this equivocation and to cast doubt on the integrity of the objectors. In other words, by trying to anticipate the likely outcome of the inquiry, objectors must offer hostages to fortune and thus put at risk their claims to impartiality.

Thirdly, aspects of the process of doing the consultancy work may themselves operate to undermine the consultants' impartiality. For one thing, having a contract from a developer allows the environmental group to direct more resources to the analysis of that developer's site and plan than to their rivals. In this case the Trust collectively had greater knowledge of C's site than of B's or A's. (Indeed, one attraction of consultancy work, described to me by a conservationist from outside

Northern Ireland, is that it allows one to 'get paid by developers for surveying a development site and then [to] use the information to oppose them if necessary. You have to be careful with this one!'; quite.) While familiarity famously leads to contempt, it can also promote a more sympathetic and rounded view of a developer's intention. In this case, detailed work on the proposed marina site indicated that this specific stretch of the ASSI contained little of high biological value, the area having already been adversely affected by long-established moorings. Accordingly, close familiarity with the site inclined the Trust to view the developer's plans as reasonable. It is at least conceivable that, had they had such close knowledge of the other developments, the UWT staff might have moderated their opposition to them also.

A fourth and closely related aspect of the consultancy process itself is that any survey is bound to be partial, in the sense of being incomplete. The UWC's survey was explicitly so; they expressly described it as 'preliminary'. Given the time constraints, which are likely to be associated with any consultancy survey, the analysis will always in principle be incomplete and thus open to question. Biological systems such as Strangford Lough are of immense complexity; definitive studies of such systems would probably be beyond any consultancy group. Biological processes are often very slow, a factor which contrasts with the speed with which consultancies must be undertaken. When environmental charities seek to do such work they may encounter particularly severe limitations since they lack the resources (library facilities, computing services, staffing and so on) of many of their competitors and, in any case, have to regard their own nature conservation work as a priority (see Cramer 1987, 49–51; Yearley 1992a, 126–40). Given all these conceivable limitations, it is a straightforward matter for opponents to make the work's incompleteness appear tendentious. To make the obvious play on words, a partial study can be presented as a non-impartial study.

Finally – and most decisively – suppose we accept that the Trust's conservationists honestly believed C to be the least bad, they would therefore naturally be eager to do the consultancy work for that firm rather than for one of the others. However, once they enter into contract with company C it becomes impossible to prove that their support for the firm's proposals is the *cause* and not the *effect* of their commercial relationship with the company. Even though green consultants acting in good faith will be attracted to working for those developers they see as least environmentally damaging, they have no independent proof that it was the comparative greenness of the scheme that attracted them – rather than, say, the prospect of a fat payment.

Conclusion

Scientific expertise is vitally important to environmental and conservation groups. It may even offer them an important means of earning revenue. But when such groups attempt to market scientific skills in situations of dispute or in adversarial contexts, weaknesses appear in the impartiality of their scientific testimony. Displays of impartiality seem to require that experts or witnesses separate judgements from factual claims; 'failure' to make this separation results in apparent weaknesses. Using one case study, I have indicated that this 'failure' stems in part from generic features of scientific knowledge, features commonly identified in the science studies literature. But the 'failure' is also encouraged by features of the adversarial context itself since barristers aim precisely to show up their opponents' experts as partial and tendentious.

Environmental organisations find themselves in a difficult situation. With a wealth of scientific ability and a growing market for environmental advice, they are tempted to sell their skills. In particular, they will want to sell their skills in relation to environmentally important developments (such as this marina in a unique ASSI) to ensure that good, nature-friendly environmental work is undertaken. Equally, however, they will want to lodge objections and to act as pressure groups in these cases also. This is a sharp dilemma. In such cases, good skills won't always yield good deals.

7
Modelling the Environment: Participation, Trust and Legitimacy in Urban Air-Quality Models

Introduction: modelling and three 'theorems' from the public understanding of science[1]

Many of the aspects of expert knowledge that today touch the public most deeply involve some aspect of modelling, whether that be of the dispersion of pollutants, the spread of infection or projections of the impact of currency harmonisation within the European Union. Though models can be of various sorts, it is clear that more and more of these models are run on computers. In turn, this development may raise new obstacles to public acceptance of, and participation in, the modelling exercises. For example, perceived difficulties with the public's ability to grasp technical issues may be aggravated because of limited physical access to the computer models or because the assumptions underlying the model are 'buried' within the models themselves. The availability of increasing computer power at declining cost makes the possibility of modelling all the greater; recently local authorities and other regional executive bodies as well as lobby groups have been able to join governments and leading research agencies in carrying out their own modelling activities. At the same time, the increased availability of computing power could be offered as grounds for anticipating the heightened democratisation (or at least accountability) of modelling precisely because 'consumer' groups may be able to offer their competing modelled knowledge. Accordingly, this case-study analysis is offered as an indication of the value of examining the use of environmental models from a 'public understanding of science' (PUS) perspective (see also Shackley 1997).

In order to set a context for this examination of the public understanding of models, I shall first set out to offer a novel systematic presentation of three key findings from the PUS literature before using them as a lens through which to view this case study.

Many of the best-known qualitative PUS studies have dealt with cases in which the public's interest and the interests of the agencies deploying the official scientific talent have been (at least potentially) at odds. Thus, Irwin *et al.* (1996) studied the case of residents living close to potentially hazardous factories. Though many residents no doubt had an interest in the commercial well-being of the factory (because they or their relatives had jobs there or because their business thrived on trade with the factory's employees) they also had an interest in ensuring that the plant was operated with as little prejudice to their health as possible. By contrast, factory mangers characteristically had a contrasting balance of interests, judging the costs and benefits of investments in safety differently. The technical staff who had the fullest access to company information and who had a full-time concern over safety-related knowledge were precisely those employed by the company. Accordingly, the question of the public's responses to the scientific issues involved could not realistically be considered in the terms customarily associated with the word 'understanding'. Members of the public formed an assessment of the scientific details of the plant's safety regime in the light of not only what they 'understood' about the technical information they were given or could acquire, but of how they evaluated the trustworthiness and the agenda of the technical staff. On the basis of studies such as this, it is argued that in the cases where science matters most to the public, the public's understanding of science is not so much a question of whether people *understand* pieces of science as a matter of the public's evaluation of the institutions of science with which people are confronted. Trust in scientists and scientific institutions turns out to be central.[2]

A second theme emerging from previous case-study analyses is that publics commonly have their own knowledges too, knowledges which may complement or rival expert conceptions of the matter in hand. This point can be illustrated by using one of the best-known case studies in this literature, Wynne's analysis of sheep farming in Cumbria (north-west England) after fall-out from the Chernobyl explosion (1992). Rainfall, which coincided with the passing of the remnants of the radioactive plume, caused upland sheep-farming areas to become contaminated. Against official expectations this radioactive contamination persisted in the vegetation and in the flocks which fed on it (see also Wynne 1994). Wynne relates how proposals to deal with this problem offered by official scientists neglected to take account of expertise which the sheep-farming community believed it possessed. For example, one suggestion made by the visiting scientists was that changing the acidity of the soil would encourage the radioactive material to become trapped in the ground and thus progressively remove it from the vegetation and, therefore, the sheep. They proposed to test this procedure by running a series of experiments in which sheep were penned in restricted areas while the soil-conditioning chemical was applied at various concentrations. Farmers objected that sheep did not thrive if penned and Wynne relates that the experiments were called off as inconclusive since the condition of the sheep declined on account of their confinement. In this case it is the deafness of recognised scientists to the knowledge of others, and not the public's problematic understanding of science, which is the core issue in the public's relationship to scientific expertise. A similar point arose in relation to officials' suggestions that the sheep be purged of the radiation by moving them onto lower-lying pastures which had not been contaminated. Farmers insisted that this was no solution at all since summer-feeding on the lower pastures would exhaust the resource needed to sustain the flocks over winter.

The third theme to arise from several of the studies develops out of the last point above and can helpfully be illustrated by the use of a well-known dispute over the safety of farm chemicals (see Irwin 1989 and, for further commentary, 1995, 111–15). Many farmers and farm labourers repeatedly expose themselves to chemicals intended to be harmful to living organisms, principally various insecticides and pesticides. Farm workers' organisations have treated occupational health as one of their principal points of campaigning and bargaining with employers' groups. Before recent moves towards European harmonisation of such matters, British regulations were decided by an expert body of scientists (the Advisory Committee on Pesticides), with members selected for their expertise in the medical effects of agrochemical exposure. This body set limits for occupational exposure, limits which were challenged by workers' representatives on the grounds that the exposure standards made fallacious assumptions about the conditions under which the agrochemical applications would be used. Clearly, while the health effects relate to the exposure of the individual worker, the regulations govern the concentrations and quantities of the preparations to be used. The link between the amount of agrochemicals to be applied and the amount with which the labourer would come into contact can only be estimated in the light of supplementary assumptions about the quality of protective clothing provided, about how disciplined workers are (or are able to be) in using full protective gear, and about other circumstantial factors such as how workers respond to adverse atmospheric conditions, such as unusually high wind speeds, or to the use of mixtures of farm chemicals. Reviewing such debates, analysts of PUS contend that such processes involve a 'naive sociology'. The regulatory body's decisions make practical sense only in the light of assumptions about worker behaviour, yet these assumptions are rarely opened to empirical examination. Similarly, technical analyses of the consequences of dumping wastes, including domestic and industrial materials, make assumptions about the conduct of the people engaged in the disposal business, assumptions which are much less often checked and based on weaker empirical understandings than the more technical aspects of the process. For instance, members of the public who live close to polluting dumpsites commonly assert that unregulated dumping takes place at night or at other times when inspection is difficult (of many possible sources, see Allen and Jones 1990, 212–13). In this sense, as Irwin and Wynne have recently commented, 'science offers a framework which is unavoidably social as well as technical since in public domains scientific knowledge embodies implicit models or assumptions about the social world' (1996, 2–3).

In sum, therefore, case studies in PUS suggest that in circumstances where the public is deeply affected by the application of scientific understanding to issues of public concern, alongside questions about the public's understanding of the relevant scientific ideas, there are questions about the public interpretation of the institutional role of science, about scientists' understanding of the public's knowledge, and about the social and sociological assumptions which underlie experts' claimed understandings. This threefold claim is represented in summary form in Table 7.1. Although I have associated these claims with particular case studies, they are borne out more broadly. Thus, medical cases can be used to support all three of the above points. From the case of HIV/AIDS research, one finds studies indicating the extent to which patients' willingness to engage in trials for novel treatments depends not only on their understanding of the treatment but on assessments of the supposed agenda of medical researchers (see Epstein 1995). Secondly, patients' groups – notably women's organisations concerned with childbirth – have successfully made the argument that the patient often has a more acute understanding of aspects of the condition than the medical professional (Entwistle *et al.* 1998). Finally, work on medically prescribed diets indicates how doctors' recommendations depend on assumptions about lifestyles and about the domestic divisions of responsibility for food purchasing and preparation, assumptions which are not subject to the same empirical scrutiny as the more technical aspects of the dietary regimen. Informal evidence suggests that diets for male patients may be used as

Table 7.1 Three theorems about the public's understanding of scientific expertise

Public understanding of science 'theorems'
1. The public's understanding of science is not so much a question of whether people *understand* pieces of science as a matter of the public's evaluation of the institutions of science with which they have to deal;
2. Publics commonly have their own knowledges too, knowledges which may complement or rival expert conceptions of the matter in hand;
3. 'Technical' understandings of science in public typically trade on a tacit or naive sociology since, in public domains, scientific knowledge embodies implicit models or assumptions about the social world.

a basis for the whole household's eating pattern whereas women patients are less likely to have the household diet altered to accommodate their needs. Commonly, this makes it harder for female patients than for their male counterparts to comply with the medical advice for sustained periods (Rose, personal communication; see also Lambert and Rose 1996).

Having outlined these three generalisations from the PUS literature, let us now consider what is known from other sources about public responses to modelled knowledge.

Models, participation and public policy

Although models have been little studied from the perspective of PUS, they have recently received a good deal of related attention as a result of innovative ideas in the field of public policy. Arguments in favour of public participation in decision-making, which had been made from a variety of sources over the preceding twenty years (for an overview see Laird 1993; Sclove 1995), were consolidated by the 'Local Agenda 21 process' which resulted from the Rio Earth Summit in 1992. *Agenda 21* was the document listing the undertakings arising from the summit (see United Nations 1993). It was divided into chapters dealing with the particular responsibilities of various kinds of agency. Chapter 28 dealt with the obligations of local authorities. Among other proposals, they were enjoined to plan for sustainable development in their regions through novel processes of participation. During the mid-1990s a series of studies was carried out looking at how such local participation might work in relation to a variety of areas, including – of particular relevance here – technical environmental issues. It was suggested that public participation in such activities might enhance both the legitimacy of the outcomes (if people 'owned' the solution, they were more likely to feel bound by it) and the quality of the knowledge. It was argued that the complexity and open-endedness of environmental problems meant that no single corps of experts was likely to be able to claim exhaustive knowledge of any system large enough to be of practical significance. For that reason, lay people who are knowledgeable because of their daily experience of their local environment might be able to act as additional peer reviewers, boosting the quality assurance of expert knowledge (an argument most famously made by Funtowicz and Ravetz 1991).

These arguments in favour of participation in the production of knowledge for environmental-policy purposes were increasingly deployed in relation to models. Perhaps the most well-known study, funded by the European Commission, was entitled Ulysses (see Pereira, Ângela Guimarães *et al.* 1999; Kasemir *et al.* 2003). It used, among other techniques, a variety of computer-based projections of regional CO_2 emissions as a tool for encouraging local debate about sustainable lifestyles and the quality of life, as a first step in facilitating local participation in such matters. Carried out in up to seven European cities and using computer models which varied from simple spreadsheets totalling a city's emissions to more complex models that tried to map spatial contributions and impacts, these studies depended on intensive focus-group discussions with randomly selected citizen panels.[3] Though not explicitly set up as studies in PUS, these studies inevitably touched on the citizens' interpretation of climate change and of the importance of changes in styles of living. The study also contributed to the PUS literature by examining the use of various techniques and technologies for enhancing group learning, by – for example – allowing groups to make graphical representations of their assessments of the impacts of changing emission levels and permitting them to transmit their views to other focus groups in the research network (the techniques are described in detail in Kasemir *et al.* 2003).

However, these studies did not thematise the issue of PUS and have not, therefore, explicitly examined the public's understanding of modelled knowledge. More significantly, these studies have been conducted on models constructed for (or sometimes with) the focus groups; they have not therefore been studies of 'live' models, that is models actually in use for public-policy or climate-prediction purposes. They thus retain a somewhat experimental or hypothetical quality.

Background to and methodology of the study

The study which supplies the empirical material for this paper was designed as an investigation into citizens' interpretation and understanding of knowledge-making about air pollution issues in Sheffield, an urban centre in northern England which employs a computerised air-quality model. Sheffield was selected as the site for the case study as it is a large-scale urban community with acknowledged air pollution problems, where the local authority was a pioneer in investing in an air-quality modelling package (the 'Indic Airviro Air Quality Management System', supplied by the Swedish company Indic AB). This package is designed to provide real-time information about air-quality conditions, to make predictions about air pollution 'black spots', and to assist in urban planning and traffic management so as to avoid and/or mitigate air pollution problems. The system, in operation since 1994, has been managed by the Environmental Protection Unit of Sheffield City Council's Public and Environmental Health Department.[4]

Information is fed into the computer model from databases (data concerning, for example, the behaviour of molecules in the air and information about traffic patterns) and from five automated monitoring stations (not all measuring the same variables). It is used not only for internal purposes – for checking compliance with national air-quality standards, for investigating connections between pollution and urban health, and for contributing to planning decisions for example – but for the production of local air-quality bulletins and the 'Air Check' system, under which air-quality information is relayed through Radio Sheffield. The information is shared with the authorities in the smaller neighbouring cities of Doncaster and Rotherham. In the recent past, Sheffield and Rotherham have established 'SARAQMI' – the Sheffield and Rotherham Air Quality Management Initiative – which aims to 'manage and improve air quality in the area' (Elleker 1995).

The research was conducted using group interviews akin to focus groups, a well established research methodology which has been applied and developed within social and market research for more than 30 years (see Merton *et al.* 1956; Morgan 1988, 1993). As is well known, the aim of the focus group is to allow the participants to develop and display their own understandings and definitions centred around core themes introduced by the facilitator. The facilitator may ask questions to encourage group discussion, but in general his/her input is secondary to that of the invited participants. Analyses of the subject under discussion are formed by the individuals in the group and through their interaction with each other. The discussion is recorded, and transcripts form the basic material for analysis.

In many other studies, this approach is further justified by the argument that participants will speak more freely in an anonymous setting. However, initial contacts indicated that sampling in communities in Sheffield would not necessarily have yielded groups of people who had neither met nor heard of each other. On the other hand, established patterns of social interaction can inform group dynamics and, in cases like this, where the research focus is on how information is processed in communities and other 'live' groups, such phenomena are not obstacles to data gathering but rather the very subject of research. Given that individuals do not form their attitudes in splendid isolation, but with regard to practices embedded in social networks of friends, professionals, relatives and so on, focus groups with existing or self-identifying groups represent an opportunity to investigate the outlooks, values and attitudes which, taken together, help to create different knowledges (see Forrester 1999). The six types of group examined in the project are: (1) environmentalists and conservationists who are members of the city's environment forum; (2) representatives of a community based around a housing co-operative located near the city centre; (3) a community group in the Tinsley area of Sheffield near busy roads and other major sources of air pollution; (4) a group consisting of public-sector workers such as health specialists; (5) representatives of local environmentalists (especially members of Friends of the Earth (FoE)) associated with traffic campaigning; and (6) a mixed group made up of people from private businesses in Sheffield.

Though there were minor variations in the way the groups were run, depending on their size, the time available and the venue, all have been based around the anti-clockwise circular format shown in Figure 7.1. Discussion started (at the bottom) with what is known of air pollution issues and the Council's computer system. This was followed by discussion of air-quality measurement. The group's attention then progressed to the processes of computer modelling, before returning to issues of public information and participation. The groups were shown screens or print-outs both from the monitoring stations (including, as an

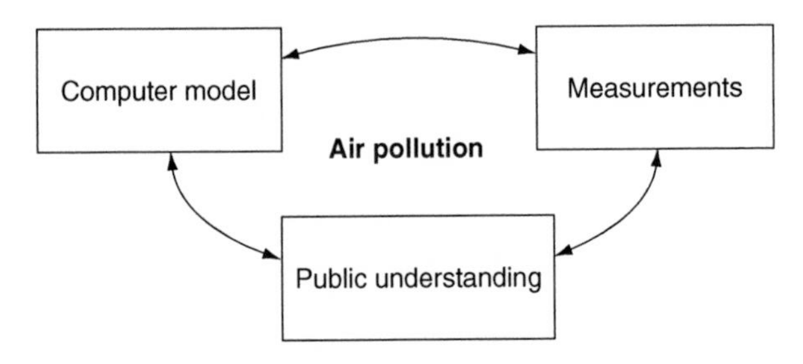

Figure 7.1 The three topics chosen to focus citizen discussions on air quality.

example, readings of nitrogen dioxide (NO_2) levels for the preceding calendar year) and from the model, usually a map of estimated NO_2 levels for the city region as a whole on a particular winter morning.

An important distinguishing feature of this study is that it dealt with an extant model in operation in a 'live' public setting. As noted above, it has recently become clear that the study of the use of computer models in public-policy settings is a key site for analysing the relations between official expertise and public understandings. Studies such as Ulysses recognise this point but because the models they examine are not in public use, issues of the public's trust and of the perceived agenda of official agencies cannot be analysed in context. This example offers an appropriate case study because, like other public-policy models, the Sheffield Airviro model is the 'property' of official agencies and therefore has not been designed with public participation in mind. Thus, though the models are built for policy purposes, the public's stake in the policy arena is not usually reflected in the model's design. Moreover, as already noted, models of complex systems necessarily contain simplifications and assumptions. The built-in assumptions in this case have not been selected by the analyst but by the model-builders and by public officials. Since it is likely that some, at least, of these assumptions relate to matters on which members of the public and campaign organisations quite reasonably hold views, it is important to study a naturally occurring instance.

Findings

Though, as will be seen, there are findings related to specific aspects of modelling, and indeed to specifics of this model (see also Yearley 2000; Yearley *et al.* 2001), it will none the less be helpful to present the results in relation to the three PUS 'theorems' described at the outset and summarised in Table 7.1.

Understandings of the institutional role of science in local governance and policy

A key issue raised by the interview groups, in some cases as their first concern, was a query about the purpose and orientation of the model. Was the model to be used principally for scientific purposes? Insofar as this was perceived to be the case, the model commonly drew criticism for being remote and impractical. The development of scientific understanding for knowledge's sake was presented as an expense which, at least in principle, drew money from other, more useful undertakings. Thus:

R1: [we'd welcome] anything which lets you say look, for Christ's sake we've got to start intervening in Tinsley. It's the stuff like other people mentioned about having an adequate factory inspectorate to monitor the steel producers and all this... All kinds of things, all kinds of funny tricks [local companies have] got to dump the waste and all like that. And it's that way, that at end of day, if you were to say to me, 'well shall we have this data, or this model, and it will cost £120,000, but being honest, we're not sure will anybody take a blind bit of notice of it? Or do you want four health visitors or four teachers, new teachers at the community school?' I'd plump for health visitors and teachers. Is it a waste of time or is it going to be effective? [3, 13].[5]

A second, and related, concern was voiced about whether the information about pollution featured in the measurement and modelling was selected for its local salience in the first place or whether the fundamental aim was compliance with national monitoring requirements:

R2: It might be interesting... I came back by car from the place I was today and I remember coming around by the ring-road on the way back and the ring-road itself would be a far better place to monitor levels in my opinion. Partly because there are people who live near there whereas the number of people who live near Charter Square, it's fairly... I mean there's a classic place... that's off the ring-road where you have a new community which has literally just been rebuilt and then there's Broom Hill which is old houses, and the ring-road goes between them. I mean it's a barrier in a lot of ways... but cars and the result of the cars around there... [5, 14].

The issue being raised is whether monitoring is being done to give some impression of what the air quality is like in the most public, city-centre locations or whether it reflects pollution levels where one might anticipate that air contamination is at its worst.

On the other hand, respondents who took the air-quality assessment system as primarily practical in orientation quickly raised questions about how valuable it would be to the public or how trustworthy or public-spirited the official users would be. The practical uselessness of the information was commonly raised. In the following extract the participants are discussing the public forecasts which are generated from the monitoring data and from the model:

Int: And are you aware of anybody, I mean so you hear [the radio broadcasts] and you think, well, dear oh dear, it's only fair today or whatever... Is there anything you can do with that information. I mean like if they say it's going to be showery – you can take an umbrella but is there anything you can do when you get that kind of warning or...

R1: Not a great deal. If you're asthmatic you stay indoors, I suppose that's about the size of it really.

Int: Mmm. Right.

R1: You can't really do a lot with it.

R2: I mean that would be my perception.

Int: Yea, no, sorry I'm not, em, I'm not trying to put you on the spot but I'm just interested that I mean you know like the pollen count you know is a similar kind of thing.

R1: Yea, yes.

Int: You know, the pollen is going to be high and therefore, you know, you want to take your, em, anti-histamine or whatever. Em, but so it doesn't come with a 'so you might like to x or y' or...

R2: No, no. It doesn't have anything like that.

Int: or wear a scarf or total body suit or

R3: [laughs]

R1: No [5: 6].

Even if the model is not seen as 'merely' scientific, its lack of practical value is taken into account in the public assessment.

A second point leading to critical assessment was that monitoring stations sometimes failed to function. According to some respondents, such malfunctions were timed to coincide with bad pollution events. When asked if it would be better for purposes of public understanding for the residents to receive information direct from the monitoring stations or a (more comprehensible) digest from the council, the answer was 'not the council!':

R3: Somebody independent because the council don't give you the true figures, as you've realised. They switch it [the monitoring facility] on and off. So that's kind of the policy isn't it [3a, 17].

The modelled information was thus being assessed in the light of assumptions about the local authority's assumed agenda. A related point, showing how understanding of the model is undertaken in a holistic manner, was made forcefully by someone working in the local education sector who had developed a special interest in the health effects of air pollution:

R2: The only two pollutants that are measured in Tinsley are SO_x [oxides of sulphur, primarily sulphur dioxide] and NO_x [*sic*]. Out of the five monitoring stations around the city, Tinsley which is under the motorway, where PM_{10} [very small particles] emissions are presumably the most high, sort of thing, levels are not measured. The pollutants from [Local Metals] in particular are a danger in terms of breathing. I was over at [Local Metals] five minutes ago, and [Local Metals] are boasting about a new filtration system which is removing far greater quantities of heavy metals that have been in the discharges. Now these discharges presumably, before the new filtration system, were much greater... From this guide to pollutants {indicates copy he's brought with him}, that's 1997, each of the heavy metal contaminants released into the atmosphere causes breathing difficulties. Now the local authority is definitely *not* measuring those things, it's definitely *not* measuring PM_{10} emissions at the local station. There's a possibility of a synergistic effect between the PM_{10} emissions from diesels and the heavy metals contamination, neither of which have been measured. Now these things in themselves you might just ignore, when you look at all the other monitoring stations in Sheffield, they are measuring at least for PM_{10}... other sites are. I can't see the sense in producing a model that's not measuring in the likeliest spots of danger to the public in that area. Now who's made the decision about what should be monitored when the same substance is being measured in other areas of the city... [3, 3].

In this case, the monitoring/modelling system is presented as inadequate; its apparent sophistication is conspiratorially viewed as a device for not detecting significant, but difficult to remove, contaminants in the very areas where they are most likely to be found in high concentrations.

That this is not an individual and anomalous response is indicated by a similarly tendentious view of a different pollutant offered by another group participant. This time the discussion was focused on the health problems associated with atmospheric lead, something which is neither measured nor modelled:

R5: we've been asking for a further study to be done, from, because it's going back nearly ten years, that what they've said to us now is well there's a lot more lead-free petrol about so therefore [air-borne lead] can't possibly be as high and that's it. That's the only answer to that. It's got to have dropped, not increased, because there's not as much lead in petrol.

Int: Right

R5: And it's those kind of responses, it's just being pushed on one side like that.

Int: So. Despite the continuing evidence...

R6: But we don't know, we don't know now, we don't know if it's continuing evidence or anything.

Int: Because they say it's not worth checking because we know up to this...

R5: Because it could have gone down because lead's gone down in petrol. We don't know it might come, that's just because they think it's come from the cars [3, 15].

It turns out that the history of the 'understanding' of this apparently technical matter is tied up with the history of ethnic relations within this disadvantaged community and with the perceived insensitivity of public officials:

R5: blood [lead] levels in the city, it was highest in Tinsley. And the suspicion, from environmental health officers at the time, was that it was all to do with the level of exposure to lead levels from vehicle emissions. And then you had a curious trace, then they moved on to lead piping. And eventually, am I not right, some bright spark decides that it may be to do with ethnicity. It may be to do with in some shape or form, the fact that there are a number of Asian families in the Tinsley area. And perhaps to Asian diet, and perhaps growing of coriander in the gardens, and the use of coriander on food, that it's to do with ethnicity determining high blood lead levels, nothing to do with the fact that the motorway just happens to be situated ten yards [9 m] away from the schools where every kid is exposed to it. And it all sort of, it encapsulates this way in which the victim becomes responsible for the condition under which people have got to live [3, 15].

Again, the 'understanding' of the modelling endeavour is inextricably situated within the context of past local authority–community relations.

A further issue in the assessment of the model concerned the ways in which the capabilities of the model are (or are not) used to inform the planning or health-policy analyses of the council and related public bodies. Though raised several times, this point was concisely expressed by members of the public-sector/health-worker group:

R3: I know that erm in the transport strategy . . . one of the meetings I went to where [the officer in charge] and the environmental protection unit, and they were about traffic management and there was a suggestion that perhaps air-quality monitoring might be used somehow in conjunction with traffic management and this was – I remember him saying 'This is music to my ears' you know as though this – this wasn't the normal course of events that this was a
Int: (laughs)
R3: breakthrough that someone in transport planning had kind of considered that they could do this, that they could work together so that – my sort of impression is that – that they don't
R2: I, I think it's something to do with the sort of departmental structures of the local authority [4, 25].

In all, though members of the public in the focus groups were curious about the model and exchanged ideas about how it worked, their understanding of it touched on a broader set of concerns than the constitution of the model itself. For one thing, they were concerned about whether it had a practical orientation or not, and exactly what 'practical' might mean. They were also concerned whether the model served to conceal truths about air pollution as well as revealing others. A final element of the understanding related to queries about how the Environmental Protection Unit's knowledge did or did not feed into the wider work of the council.

Expert and lay technical knowledges

The second issue concerned people's own knowledge about the 'technical' aspects of the monitoring and modelling. This was important in relation to both how the data for the model are acquired and fed into the computer *and* how the model handles these data. On the latter point, people, notably respondents from the city's environment forum, questioned how the model itself functioned: for example, where is the boundary of the model and how does it handle chemical as opposed to physical interactions between gases? Though this might be taken by modellers as primarily an issue of model resolution, respondents were not slow in offering political interpretations for the ways boundaries are drawn and for the ways in which the pollutants are selected (or not) for measurement and modelling.

With regard to the former point about data acquisition, public views about gases and other substances omitted from the monitoring programme have already been mentioned. Other points raised about the data concerned the location of the monitoring sites: whether they were in unrepresentatively quiet streets or close to bus lines which were 'known' to be cleaner than the average, and whether they measured pollutants appropriate to the situation in which they were located. On this latter point, respondents noted that there was concern in the west of the city about a proposed incinerator for cattle carcasses (required as part of the attempt to dispose of animals which might be carriers of Bovine Spongiform Encephalopathy (BSE)); none of the monitoring would help detect or regulate emissions from such a facility. Equally, the traffic campaigning interviewees (Group 5) mentioned earlier publicity activities they had undertaken relating to acid emissions from power stations about 80 kilometres north-east of the city. They therefore had independent sources of knowledge about air pollution threats to the city which the model did not appear to highlight.

Respondents also suggested that they had local forms of technical information to contribute, data which were overlooked or even spurned.

R4: I remember, I don't know whether it was [Sylvia] or it was [Joan], someone said that they cleaned out the gutters; when you took them down to get them inspected because you were concerned about what you were finding one day, environmental health said 'we're not interested in what's in your gutters'. And for me, that was the cardinal sin [3, 15].

People were taking a concerned interest in evidence about contamination which they believed had come from a local factory or other plant, but this was rejected by the council's experts. Personal knowledge was also drawn on in assessing the adequacy of the modelling. After looking at the modelled projections, one participant from the business group commented:

R1: yes my youngest was diagnosed as asthmatic
Int: d'you – d'you see any relationship between these poor are -are -are er does your -your -your child erm watch out for these
R1: no never and it's an area which was interesting is an area over which you can't impact as an individual and obviously whilst we wouldn't go out of our way, I mean neither I nor my wife smoke
Int: uh huh

R1: so therefore we become tetchy when people come into our house and smoke

Int: yes

R1: but how you relate the health of your child to the air affecting the county of South Yorkshire is – is a much more difficult question

Int: sorry?

Int2: and the other thing is ignoring this sort of information from the council, the met [meteorological] office and all over I mean do you – have you a personal assessment of the air quality then of your child?

R1: oh well your own – your personal assessment I mean it's where your model gets thrown, we – we live at S10 [a postal code indicating a relatively prosperous part of the city] right under the first tee of the golf course – now the problem is, you get low cloud round there; I don't need a measure of the air quality, when I step outside and I can smell the smell I know it's cos it's not escaped

Int2: yeah

R1: now in theory I'm sure my chart shows that I've got quite good air there; most of the time we have but I would say when it's bad it's probably poorer than most other people's cos it literally doesn't escape [6, 22].

Two points emerge from this quote. First, it is interesting to see how the respondent links the internal and external environments around his house. Indoor air contamination (from smoking) is as much a hazard for his children as outdoor atmospheric pollution, but the relationship between the two has to be handled in terms of lay knowledge since the model does not deal with the indoor environment at all. Second, his assessment is that the model has deficiencies when it comes to specific locations in the city, even if the overall pattern generated by the model appears credible. That this view is not merely idiosyncratic is suggested by the responses of the public-sector professionals group:

R2: erm we did look at the outputs of the model in connection with the transport and health work [we were engaged in]... and for me I mean the limitation of it seemed to be that it's largely based on traffic flows rather than measurements of air quality

Int: right

R2: so it's of air quality

Int: yes

R2: based on various assumptions about links to traffic flows and topography and climate and so on, meteorological

Int: yes

R2: things, so erm and I think in talking to other people who've just seen some of the pictures that's not – that's not always the perception; the perception sometimes is that these, this is an actual picture of air quality across the area

Int: yes

R2: and they're quite surprised to discover that in fact that's not the case

All: (laughter)

R2: and the air quality's only measured in a couple of places [4, 9–10].

In subsequent discussion, these respondents expressed the view that the model was so dominated by the assumptions listed in the middle of the above quote (about traffic flows and so on) that it could not be used for epidemiological or similar purposes. Health workers made no effort to correlate illness patterns with the spatial readings from the model because they believed that the model was far from sufficiently precise. This view was offered tacit support by one of the model operators who commented that, when he walked around Sheffield, his perception was that pollutants were much more broadly dispersed than the model implied (because it associated pollution so closely with transport arteries).

A related point was raised during the group interview with traffic campaigners. Many of the reforms which they believe they might be able to get the local authority to implement would be small-scale, involving the closure of small stretches of road to private vehicles or the introduction of various traffic-management plans. In relation to the monitoring, their claim was that:

R2: this comes back to where they've got their monitoring stations – there aren't enough of them to give you the information that really would provide answers to whether it *was* working [5, 42].

Finally in this section, as noted in the last-but-one quote, the model places considerable emphasis on traffic emissions and thus its predictions for each day depend on projections of traffic, themselves based on periodic vehicle surveys. Local residents viewed themselves as particularly knowledgeable about traffic trends, especially on the characteristics of irregular (though not infrequent) special occasions:

R8: there's all these night clubs, I mean many a night if Tina Turner's on or somebody, great for us, we're down the road, no transport, cos you're there, you can go to see 'em. I love [the] Arena Stadium,

I've been to all the lot, but the traffic is absolutely stood on Bawtry Road. So you've got traffic on Bawtry Road which is stood, you've got traffic on the motorway which is stood, and everything comes into junction 34. And it doesn't matter what they're planning, I mean they've made a new road for the airport but there's still a lot of it, if it takes off and it's busy, we've got 'em at junction 34 again. Footballers [i.e. supporters] coming in on Saturday, no matter whereabouts in the country they come from, most of them come up the motorway and into junction 34...

R7: And when...it's busy, like at Christmas

R8: Christmas, we always say November, with Christmas, on Saturday and Sunday...Christmas in Tinsley. I mean I go to the 'house show' in November and we always go round [the long way] instead of going this way cos you can't get out. [3, 19].

Though the statistics may be able to deal with exceptional traffic surges which are repeated (the January sales for instance), in the case of other irregular events local people feel that they have the more accurate information.

Underlying sociological assumptions

The final point in the last section (about traffic flows), as well as demonstrating local knowledge, relates closely to the third general PUS theorem concerning the role of assumptions of sociological nature. Though vehicle emissions might be regarded as a technical matter, driving behaviour could easy be categorised as social. Thus it was not only in the Tinsley groups that questions were raised about traffic data, for example about how frequently the survey data are updated:

R2: [there] is one thing...you say they have taken traffic measurements – are they static or are they [voice trails off] because the thing that I found interesting about this is they're now doing real time monitoring of the pollutants but are they doing...real time monitoring of the traffic or is the monitoring of the traffic still based on one-off yearly surveys? [5, 34].

Other respondents, particularly members of the city's environment forum, mentioned the apparently counter-commonsensical possibility that the city might suffer from air pollution caused by agricultural practices in the surrounding countryside (notably in the form of organophosphates). And the topic of indoor or other kinds of domestic air pollution, caused by residents' behaviour, has already been referred to in the quotation from the business group respondent.

However, respondents also made reference to the model's sociological assumptions in a way closely resembling the findings of the other qualitative PUS studies reviewed in the introduction. As well as using traffic data and so on, the model depends on the input of information about permitted emissions from large plants. Yet a main claim of residents concerned the covert behaviour of plant managers. One main source of information here was said to be people working in the plants:

R8: One of the things said at [the] last meeting is some of the men that work at [Local Metal] said that they actually shut down the fans or what takes all the pollutants out in the middle of the night and they just release it
R9: They do that. They do that at Rotherham, you know the one at, eh,...
R8: They do
R9: [Another Metal], they do it there, they close it down and the same happens there
R8: It's not too long since...
R7: And that's not too far away from here either
R8: ...here we got like a sulphur smell, like, just after tea time [3a, 29].

The other source of knowledge about such irregularities was said to be local observation of the patterns of odours:

R1: The steel works turn off the air-scrubbers because they're expensive to run, and just let it, you know, free float into atmosphere
R4: Have you got any evidence for that?
R1: No...but within half-an-hour it's gone...
R9: ...well that's the kind of thing that they want to know
Int: So you say you can actually see, you can actually smell...
R1: Yes
R9: I can
R1: And we don't know where it's coming from
R5: Well we do know, we've got a rough idea of wind direction...
R6: We've got people who work at these firms who will tell you that these firms that they work for pollute the atmosphere in dark, under cover of night [3, 22].[6]

This second extract appears to indicate that local people recognise how difficult it is to acquire definitive evidence of improper conduct in local factories, and they revert to information from insider sources allied to the community. However, odours do appear to be regarded as strong evidence in the community, a finding reflected on the other side of the city where the anti-cattle-incinerator campaigners have successfully mobilised local information about the spread and persistence of foul-smelling odours from an abattoir close to the proposed site.

Concluding discussion

At its simplest, this case study has indicated that key claims made in qualitative PUS studies of science in public contexts also apply to the public's response to scientific models in the public domain. Though this model was bought 'off the shelf' and then customised and operated by a public body (a local authority), the public's understandings of the model's outputs were clearly influenced by three kinds of factors: people's assessment of the trustworthiness and agenda of the council; people's confidence in the technical knowledges which they believed they already possessed; and their evaluation of the social assumptions which underlay the model.

However, this case is not simply of academic significance since the public's understanding and acceptance of the model have practical consequences. The model and its forecasts are intended to affect how citizens behave and how they respond to the local authority's policies. The council tries to disseminate information from the monitoring and modelling programme so that people can better understand and make choices about their environment. But if they are sceptical about the value of the model then they may not use it; indeed they may be tempted to regard model outputs as the sheep's clothing donned by the wolves of the planning department. Equally, the public's understanding of the modelling process as a whole may well affect their willingness to engage in future monitoring and forecasting activities undertaken by the council.

As noted in the introduction, the spread of cheap computing and the availability of commercial modelling products are hastening the spread of computer modelling to more and more aspects of public policy and personal planning. Without detailed analysis of the public's understandings of these models, supposedly generated on their behalf, it will be impossible to comprehend why models are perceived as accurate, trustworthy or legitimate. The overall finding of this case study is that publics both believe themselves to be substantively knowledgeable about many of the topics modelled and assess policy models in the light of a broad range of social and political factors. One conclusion following from this finding is that to build robust and legitimate models, public bodies will need to devise methods of consultation and participation not only when the model is running but in setting out the objectives and parameters of the model in its earliest stages.

Part III

Cultures of Knowing and Proving: Science, Evidence and the Environment

8
Green Ambivalence about Science

Introduction: green ambivalence about science

For its twenty-first 'birthday' in 1992, Friends of the Earth (FoE) (England, Wales and Northern Ireland) published a celebratory booklet. With a large supporter base, regular coverage in influential media, strong campaign teams and widespread recognition of its name, the organisation had a lot to celebrate. Yet the item chosen to begin this celebratory publication, immediately after the contents page, was a quote from a leading environmental journalist praising the group as a 'reliable and indispensable source of information'; this was followed by a comment from the head of Her Majesty's Inspectorate of Pollution (the forerunner to the UK's Environment Agency) praising the quality of its 'technical dialogue' (1992, 2). Of all the items which could have been chosen to feature on the second page, this selection was surprising and telling.

What it told was of a long move from stunts and publicity-seeking to a more sober style of campaigning. In the quarter century since they were set up in Britain, groups such as Greenpeace and FoE had moved towards an embrace of science and 'technical dialogue'. The adoption of science allowed them to become more effective in their role as 'insider' pressure groups and has won them respect, widespread coverage for their stories and some notable improvement in their policy prescriptions. At the same time, it has consequences for how the organisations are run and for the way they approach campaigning. Of late such groups have sometimes been accused by community-based activists and by journalists of becoming bureaucratic and inflexible (see Allen 1992, 220–23 and Chapter 2). By contrast, I believe that their accommodation to scientific expertise is a large part of the explanation for their changed 'body language' (see Yearley 1993).

Science has been a key element in the UK environmental movement all along. The leading nature conservation organisations in Britain have a long history of granting science a central place in their activities. Thus, the forerunner of the Wildlife Trusts, the Society for the Promotion of Nature Reserves, actually argued for the conservation of nature primarily because it permitted the pursuit of biological research. This emphasis is reflected in the most common current designation for protected habitats, Sites of Special Scientific Interest (SSSI) (as described in Chapters 5 and 6). The key development has been that in the last twenty years the younger environmental groups have tended to converge with such conservation organisations by building their own scientific staffs. Greenpeace now happily boasts of its own laboratory facilities, and more care is taken to get reports and publications reviewed by 'peers', technically proficient personnel outside the organisation. In fact, this convergence has been two-way, with the more traditional groups too recognising the need for political pressure, and sometimes recruiting staff from the more radical groups to run their campaigns.

It seems we might join with the distinguished naturalist Nicholson who claimed that conservation 'should be science led and science based' (1987, 81). But there are two important grounds for being hesitant about such claims. For one thing, many of our environmental problems originate from the scientific and technological nature of our civilisation. In many instances the connection is clear and direct. Humans invented the CFCs which threaten the ozone layer. Technological advance allowed humans to develop nuclear power, which in turn has brought us persistent environmental problems, such as those associated with the calamitous reactor explosion at Chernobyl in the Ukraine in 1986. As Beck has insisted, in such cases we can trace environmental problems directly to specific products of science and technology (1992). There is also a more diffuse connection: present-day industrial society is inseparable from pollution caused by motor vehicles, power generation and waste disposal. Many environmentalists are thus critical of technical progress and, at least, ambivalent about science.

The second reason why environmentalists are loathe to grant science a leading role is that, on closer inspection, scientific expertise soon begins to lose its straightforward appeal. In many disputes over environmental policy, scientists are aligned on both sides; for instance, some highlight while others play down the problems associated with toxic waste materials. And, although the nuclear industry depends on very high levels of scientific skill and draws support from many eminent scientists, Rüdig asserts that dissenting scientists were particularly influential in the anti-nuclear movement (1986). The industry has its scientific critics as well as its scientific advocates. As we shall see, there can be various reasons for disagreements among scientific experts. Sometimes, the relevant evidence is simply hard to obtain or there may be a range of scientific information which has to be taken into account. In such cases, scientists would not necessarily be expected to agree. On other occasions, lay persons may suspect that the scientists associated with a corporation or industry are not acting in a truly impartial way.

To find that scientific evidence, far from being an impartial resource for resolving a dispute, may become part of a pressure group's campaign armoury leads us towards the 'social construction of social problems' approach advocated by Kitsuse and Spector (1981, see also Hannigan 1995; Yearley 1992a). Such analysts of social problems suggest that the correctness of social problem claims may be comparatively unimportant in determining their public impact. If scientific truths can be matters of dispute, particularly among experts, then even support from science may afford only a limited impact for those making claims about environmental social problems. If we accept this point then it is harder to view the ecological movement as science-led and science-based in any simple way. We can usefully begin our investigation of the peculiar scientific dependence of the ecological case by comparing environmentalists' arguments with those of other campaigning groups.

The green case and scientific authority

Despite the fact that, unlike many preceding social movements, the environmental movement claims a scientific basis, little social scientific attention has been focused on the particular role played by its scientific credentials. Such a claimed basis is not unique; movements for the adoption of scientific medicine and for public hygiene, for example, shared it. And, of course, the label 'scientific' is itself open to negotiation and extension, so that movements dedicated to preparing a welcome for visitors from outer space or to the promotion of a biblically based Creation Science might be heralded as scientific, at least by their supporters (Wallis 1985). But the ecological case is very profoundly a scientific case. Amongst its strongest planks is the argument that disaster for the human race is a natural inevitability if certain practices continue: if we persist in allowing CFCs to escape into the atmosphere, if we cut further into the world's rainforests and if we do not reduce the emission of greenhouse gases. Other arguments concern not our survival or even the perpetuation of our current standard of living. They concern the damage to the world's other inhabitants: the need to protect wildlife by not bleaching paper or nappies with chlorine, by conserving endangered habitats and by not using detergents that are rich in phosphatic water softeners.

While each of these arguments is now beginning to command public attention, it is important to appreciate the extent to which they are all based in a distinctively scientific perception of the world. Thus, to take the most dramatic example, the ozone layer is only available as an object of knowledge because of our scientific culture. At ground level, ozone is relatively uncommon and remote from experience. The stratosphere where it is prevalent is, if anything, more remote. Knowledge about the hole in the ozone layer is only available through high-technology ventures into the atmosphere high over the poles. Equally, our everyday supposition is that detergents have done their job once they have left our washing machines or sinks. We would not readily think of the damage they might be doing in water courses. It is biological scientists who detail the connection between water-softening agents and the algal blooms which choke life out of the water courses. Finally, there are examples where environmental damage is done by massed products of which each of us uses only a little. It takes an unusual degree of determination to total the leachate from apparently innocuous rubbish tips or to work out that many millions of tonnes of carbon dioxide (CO_2) can be pumped out from our car exhausts. Ordinary members of our culture would have no real conception of what even a tonne of gas was like.

Thus, the environmental case is tied to science. It states that natural realities constrain our options in various ways. Moreover, these constraints are relatively independent of people's moral standpoint. By this I mean to contrast the greens' arguments with the motivating concerns of other social movements such as those involved in the dispute about abortion and the right to life. Whilst 'life' itself has the appearance of a scientific notion, and scientific evidence is brought to bear on the debate (in terms, for example, of when the various bodily organs in the foetus start to function), scientists are generally reluctant to pronounce on the point at which life begins. They commonly present this as a metaphysical rather than a scientific issue. Additional moral arguments are also brought to bear in the abortion case; for example, it is often argued that, if 'pro-lifers' are serious in their convictions, then they should be campaigning with equal vigour for more help for AIDS sufferers or for those in developing countries who are struggling with life-threatening problems.

To take another example, we can turn to the case of campaigns against 'too much' sex and violence on television. The claims about exposure to too much sex are typically made in moral terms. Freedom of artistic expression is weighed against moral propriety. Of course there is an empirical aspect to this dispute. Campaigners may assert that as a matter of fact 'overexposure' to licentious material leads to moral degeneracy. But such claims have proved very resistant to empirical test; degeneracy, for example, is itself a morally charged notion and thus hard to measure in an agreed way. Furthermore, those interested in the debate often call on their personal experiences to assess the validity of the competing 'factual' claims and are reluctant to hand the matter over to those, such as psychologists or media analysts, who claim to have disinterested expertise relating to these facts.

By contrast with these social problem issues, environmentalists would see their case as unanswerable by virtue of its scientific credentials. The ozone layer *has been* disappearing and the consequence *will be* greater amounts of damaging radiation; greenhouse gases *are* accumulating and a consequence *will be* the expansion of oceanic waters, the melting of land-based polar ice and the associated coastal flooding. These are held to be matters of fact and there is little room for moral dispute about them.

Many conservationists see their case as motivated, in Moore's words (1987, xviii), by 'objective reasoning'. Moore, who was a leading scientific authority in the Nature Conservancy/NCC until his retirement around two decades ago, relates how he was able to establish that toxic pesticides accumulated in the food chain (1987, 157–66). It was not a question of opinion; he had demonstrated that through unexpected but relentless natural processes, farming practices were threatening to poison people. The scientific demonstrability of the connection between the use of agrochemicals and the presence of toxins in food was decisive. Max Nicholson makes the same point. Again speaking of the organochlorine insecticides, though this time in relation to attempts to persuade the agrochemicals industry to change its policy, he states: 'Had not the scientific base of ecology and conservation been already so sound, the successful agreement with the industry could not have been concluded' (1987, 49).

Moore and Nicholson suggest that the scientific credentials of the conservation movement lend it considerable authority. This claim can usefully be understood in relation to Weber's famed analysis of the social bases of authority. Weber identified three kinds of authority: the traditional, the charismatic and the legal-rational (1964, 328–29). In his view, claims to authority may be made on the basis of hallowed and accepted wisdom. Or they may be made on the grounds of the personal authority of a special leader, deity or prophet. Or they may be made impersonally, on the basis of accepted and demonstrable principles. Weber argued that the last of these, legal-rational authority, increasingly pervades contemporary society. According to scientific conservationists like Moore and Nicholson, it appears that the environmental movement is peculiar in its commitment to such authority. Other social and political movements commonly draw on traditional authority (for example, nationalism), charismatic authority (new spiritual movements) or on some combination of these.

Moreover, as both Nicholson and Moore acknowledge, the environmental movement's ties to science extend from the present into the past. Many of the more 'establishment' nature conservation societies have a background in natural history: the Royal Society for Nature Conservation (RSNC), the British Trust for Ornithology, the various Naturalists Field Clubs and so on (see Sheail 1976, 48–53, 1987). Scientific views and scientific authority have been central to their organisational development, as Nicholson remarks (1987, 91–92):

> While parallel movements such as those for development and the relief of poverty in the Third World, or for peace and disarmament, have had to create their own often sketchy foundations, environmental conservation differs in resting upon a comprehensive and profound set prepared for it in advance by the natural history movement.

On this view the environmental movement is doubly bound to science, by epistemological affinity and common descent.

Legal-rational authority and the sociology of science

The special authority enjoyed by legal-rational forms of argument has often been taken for granted by social scientists. Indeed, since social science too appears to rest on this form of authority, to throw it into question might seem a self-destructive pursuit. But in the last two decades there has been a reassessment of this kind of authority among sociologists and philosophers of science who have studied decision-making in what might be taken to be the temple of legal-rational thinking, natural science. What they have suggested is that the authority commonly associated with scientific beliefs is not as straightforward or as unequivocal as many people, including Weber, appear to have assumed (see Barnes 1985, 72–89; Yearley 2004). These sociologists and philosophers have argued that the public, policy makers and more traditional philosophers of science have exaggerated the authority of science. Essentially, there are two components to this claim. First, they argue that scientists' judgements inevitably go beyond the evidence on which they are based, so that scientific authority cannot be justified by a simple appeal to its factual foundations. Second, they argue that even the facts themselves on which science is based do not command an unquestionable authority.

Let us start with the status of facts. Our confidence in the ability of facts to validate scientific beliefs often draws support from an analogy with perception. Facts are often taken to be evident in the same way that the things we see are evident; we are passive recipients of knowledge in the same way as we receive the evidence of our eyes. However, this analogy ironically works to undermine the case which it is supposed to support since perception is far from the passive operation that this argument implies. To observe is to do far more than merely allow one's retinas to be bombarded with light since objects which are 'seen' as the same actually present very different appearances to the eye.

For example, as sunlight is reduced by a passing cloud, the colour and appearance of objects as measured, for example, by a light-meter change, yet we regard the image of objects as staying the same. Equally, we can retain a constant interpretation of an object despite approaching it from many different angles. In these ways we 'see' more than actually strikes the eye. At the same time, we use only a minute amount of the visual information potentially available at any given time. Seeing is usually the imposition of interpretative schemas onto the available information; observation is a blend of interpretation and the reception of light. The extent to which humans habitually depend on interpretation in seeing is demonstrable through such things as optical illusions and trick pictures. Just as perceptions have to be worked at, so do observations of facts. An unskilled observer allowed to wander in a wood will experience great difficulty in 'observing' the different kinds of trees for there are likely to be as many variations between members of one species as there are between the differing types. The difficulties would be magnified if the observation were done over a protracted time period so that buds came and went, fruits appeared or leaves fell. An experienced botanist on the other hand would not only immediately see the kinds of tree but would be able to observe higher-level degrees of similarity between different species; see signs of health or disease; and see whether the season was late or early. In putting the argument this way I am not trying to raise doubts about botanists' knowledge or expertise. I am not arguing that there is something deficient in their seeing. But I am aiming to throw doubt on the idea that scientific knowledge is valid simply because scientists see the world plainly. All useful seeing is skilled seeing.

But the argument does not stop there. The status of scientific facts is even more complex than this analogy with perception reveals. Many scientific observations are made with machines: one observes an electric current or radiation from space not with one's senses but with an instrument. Similarly, 'holes' in the ozone layer cannot be seen in any straightforward manner. Their existence is inferred from detector readings obtained during flights over the poles. Scientific observations, particularly in the context of experiments, also have to be separated out from chance occurrences. A rise in sea level due to freak weather conditions will look much like a rise due to global warming. Even scientific apparatus appears to be inhabited by gremlins, and observations of changes in a meter reading or of blips on a pen-chart have to be divided into real facts and mere artefacts. Furthermore, the frontier of things which count as factual observations tends to shift as scientific ideas change so that what, at one time, would have been regarded as hypothetical images from a new and experimental form of microscope for example come later to be regarded as unproblematic observations. Finally, the facts of interest to scientists are commonly not single, isolated facts but are facts about classes of things: for example, that ultraviolet radiation admitted through a hole in the ozone layer would damage plankton. In such cases it is evident that no one could have made observations on all plankton, certainly not those which are not yet living. There is, therefore, something undeniably conjectural about factual claims regarding a whole class of phenomena.

So far we have looked at scientific facts and the problems in establishing facts by observation or experiment. This focus corresponds with one facet of science; sometimes science is held up for admiration as a body of factual knowledge. Most often, though, it is theoretical knowledge which is regarded as the principal achievement of science. The expectation that the Earth's temperature will rise because of the greenhouse effect is a prediction based on scientific theory. The suggestion that algal growth is encouraged by discharges from farm and domestic waste is based on theoretical understanding in biology. Sociological studies of scientific reasoning have indicated that theoretical judgements in science are not dictated by the factual evidence in any narrow sense. In part this is a logical matter. In principle at least, a set of facts can always support more than one theoretical interpretation. But also, at a practical level, it is generally impossible for scientists to collect all the information they would like before having to decide on a theoretical interpretation. Scientists' theoretical beliefs are not fully decided by the factual evidence available to them.

I should repeat at this point that these arguments about scientific knowledge have not been introduced in order to criticise science or scientists. They have been introduced because they indicate the ways in which the acquisition of scientific knowledge is far more complex than is normally supposed. Scientific knowledge depends on judgement; therefore the environmental movement's dependence on science is itself not likely to offer the movement a straightforward appeal to authority.

The abstract points made in this section can be exemplified if we turn to the case of an experimental test which was designed to address public fears about one particular environmental issue. Although this case is not necessarily typical of the role of science in environmental matters, it is illustrative; it has been described in some detail by Collins (1988).

The UK atomic power industry collects together spent fuel from the nation's various nuclear power stations at Sellafield in Cumbria for storage and possible reprocessing.[1] The spent fuel is transported by rail in special containers mounted on flat trucks. Many environmental groups have been concerned about the hazards which could result from reprocessing and they have sought to focus attention on risks associated with the various stages in the reprocessing operation. One stage which has been highlighted has been the transport. Objectors have tried to raise doubts about the safety of the flasks in which the material is transported: what would happen if there were a collision, or a fire, or a terrorist attack?

To some extent their anxiety has been shared by local authorities through whose districts the flasks are transported. As early as 1984 these questions were being raised insistently and the state electricity company (CEGB, as it was known before privatisation) took steps to quash these doubts. They did this by arranging an experimental test. In front of 32 carefully sited cameras, a remote-controlled locomotive, pulling three carriages, smashed into a 'derailed' flask and truck at 100 miles (160 km) per hour (Collins 1988, 731–32). The steel flask was flung into the air by the force of the impact but was subsequently seen to be substantially undamaged. The measured pressure inside had fallen by only a quarter of one per cent, and had it contained nuclear material, no leak would have resulted. The Chairman of the CEGB, an enthusiast for nuclear power, was very pleased with the result and argued that the general public's doubts should now be dismissed. He suggested that while it might have been possible to harbour doubts about the flasks' safety when the public only had the CEGB's word for their safety, this empirical test should be fully convincing (1988, 726).

However, when interviewed by Collins, a Greenpeace representative claimed that he was unconvinced. It was not that he doubted that the flask had survived the collision, nor that the train had been travelling very fast. Rather, he argued that certain features of the test materials rendered the experiment unrealistic. For one thing, there are flasks of various designs in use. The one in the test had thick, solid steel walls while other models are composed of a thinner steel casing lined with lead. Such a flask, he feared, might not have withstood the blow. Moreover, he claimed that the kind of locomotive chosen for the test happened to have a softer 'nose' than others in common use; accordingly, some of the force of the impact would have been absorbed by the crumpling of the locomotive rather than being transmitted to the flask.

The test was subjected to further criticism because of the arrangement of the collision. It has already been mentioned that the flask was flung into the air in the crash. This was possible because the flask was placed at the end of a section of track with a large clear space behind it. The Greenpeace spokesperson argued that this detail made the test rather too easy for the CEGB. The impact of the crash was not sustained precisely because the flask was flung aside. Very different circumstances can easily be imagined. For example, had the flask been caught between a locomotive and a hard surface (the side of a cutting or the inside of a tunnel, say) the forces on the flask would have been more intense. Greenpeace's representative suggested that such a scenario was not far-fetched; indeed, the conditions arranged by the CEGB for the test could readily be seen as less 'lifelike'. As Collins remarks: 'Presumably Greenpeace would have wanted a harder-nosed locomotive, with differently weighted carriages, with the flask pressed against the abutment of a bridge at the moment of impact, thence falling from a high viaduct on to a hard surface and into a fierce fire' (1988, 738).

Using this example, we can return to the issue of the character of scientific knowledge. I described two important considerations which affected scientists' authority: first, that their judgements inevitably go beyond the factual evidence on which they are based and, second, that facts themselves do not enjoy an unassailable status. This case study bears out both these points.

Let us look first at the facts of the case. In the narrowest sense we can say that a specific train, travelling at a certain speed, collided with a particular flask without rupturing it. But it is not easy to progress from this rather uninformative statement. Since there are various designs of train and different types of flask, it is unclear whether we can say that flasks can withstand (high-speed) impacts from trains. The CEGB's representatives might well propose this as a 'fact', while Greenpeace spokespersons would doubt it. There is a clear photographic record of the impact but the parties continue to dispute what facts these observations support.[2]

This leads us to the issue of judgement. Even if the parties had agreed that the experiment showed that all nuclear flasks can withstand such impacts from all relevant British trains, it would still be unclear whether this collision can represent all mishaps which may befall the flasks. In other circumstances there might be an explosion or enormously high temperatures; how would the flask behave under those conditions? The question of the safety of the flask requires an act of judgement in addition to the experimental test.

As I stated earlier, there are anomalous features to this test. Scientists would never normally accept the result of just one impact. Equally, they would generally require that the flask be tested under a range of circumstances. Collins argues that although this collision looked like an experiment, it was really more of a demonstration exercise. It was designed to display the safety of the flask rather more than to test it. All the same, this case helpfully demonstrates the complexities of turning to science for the adjudication of an environmental policy problem and indicates the ways in which the authority of scientific pronouncements may be challenged. No matter how many tests there had been, Greenpeace could always have queried the 'facts' and could always have thought of novel and bizarre ways in which the flask might have been damaged but which were not covered by the experiment. They could challenge the facts and challenge the judgements which allowed general conclusions to be drawn from those facts.

Science as an unreliable friend – empirically

Up to this point I have examined in general terms the part scientific arguments play in the ecological movement, and some of the limitations which may beset scientific authority despite the respect which it normally commands. It is now time to investigate the use of science in the work of environmental campaigners. Conservation and environmental organisations do make extensive use of science; they call on expertise on, for instance, food chains, species identification or energy conservation.

Size for size, they have large numbers of scientifically trained employees. They typically have scientific advisory committees and call on the voluntary support of university scientists and scientific civil servants. As we have seen, even the more radical of them are increasingly interested in taking on scientific skills. But is this scientific base sufficient to grant environmental groups the authority to which they aspire? This question can be answered by taking a look at the ways in which their scientific skills are typically 'cashed in' and asking how good a friend science is to the ecological movement.[3]

The first way in which science is an unreliable ally is a simply empirical one. Compared to social movements that appeal to an orthodoxy or to a charismatic leader, avowedly scientific movements face a number of pragmatic disadvantages. Scientists may not have an answer to every question. Similarly, they accept in principle that their knowledge is revocable and incomplete. This incompleteness may manifest itself in a number of ways. For example, the scientific committee of one of the groups studied, the Ulster Wildlife Trust (UWT), were concerned to learn of proposals to use a quarry some thirty kilometres north of Belfast for dumping domestic refuse. At the time, the principal site for domestic waste dumping in the Belfast area was the north foreshore of Belfast Lough and there was considerable anxiety about the impact of this practice. It was most likely damaging to marine life in the lough and to visiting bird life; the site is also flat, open and clearly visible. Thus, the prospect of having an alternative site for dumping was initially attractive. However, there were also reservations about the control which might be exercised over the leachate from the proposed quarry site, especially since the quarry was close to Larne Lough, another sheltered marine lough which is the site of an RSPB (Royal Society for the Protection of Birds) reserve. The committee members had no special expertise on the matter of leaching and did not wish to appear oppositional for the sake of it, but they were uneasy about being seen to consent to the quarry-fill. The scientific committee found itself in a rather 'counterfactual' position: while they assumed that there was an answer to the question 'which site is better?', they had to acknowledge they did not know what it was.

The same problem can be even more acutely felt if there is some public pressure to announce an answer. In a subsequent case the same organisation was invited to comment on a proposed holiday development. In this instance the scientific committee of the UWT was very gratified to find that they had been consulted about plans to extend a sewerage system for a caravan park and amenity centre which happened to be close to a reserve of theirs. However, they had no specific expertise on this subject and no precise way of knowing what the effects would be. They felt uneasy about their lack of knowledge in the light of the unusual invitation to comment. To have nothing to say when specifically asked to comment seemed to contradict their claim to speak for nature's needs; but there were equal dangers to speaking without authoritative knowledge.

Although the explicit invitation to comment made this case unusual, this difficulty in responding to planning proposals is not uncommon. Often it is unclear ahead of time what the implications of a scheme will be. It is tactically unattractive to object to everything but it is hard to be seen to be condoning something which may turn out to be harmful. This position might be seen as putting the scientist in the position of spoilers who are tempted to view all change as a disturbance of the natural order and to ascribe all environmental damage to human influence. The vulnerability of environmental groups' scientists on this charge can be seen with regard to TBT (tributyl tin), an anti-fouling preparation applied to boats and other marine equipment.[4] The use of this substance, especially in important marine areas, is opposed because it is likely to be harmful to aquatic life; under test conditions it can be shown to be associated with developmental abnormalities in shellfish. In opposing a plan to site a salmon farm in a sheltered marine lough in Northern Ireland, conservationists were concerned that the fish cages would need to be treated with TBT or a similar preparation. Although plainly worried by this possibility, members of the UWT's scientific committee were aware that they could not say precisely what consequences would follow from such a use of TBT. In part this was because different uses have different effects. For example, the application of anti-fouling agents to boats distributes the chemical around the lough, while the results of using it on cages would depend much more on current directions. Again the scientists were in the awkward position of assuming that there was a scientific answer to the question of the anti-foulant's effect but of acknowledging that they did not know what it was.

There was a clear feeling in the UWT, as in other Wildlife Trusts affiliated to the RSNC, that there is a need to be 'reasonable' and to not oppose all development, yet the very rigours of scientific evidence and proof render 'reasonability' very difficult to achieve. The situation is further complicated since the conservationists have an additional reason for wanting to oppose potentially harmful measures: they consider that it is harder to stop something from continuing once it has been set up than to prevent it altogether. For instance, once jobs have been created by a fish-farming project it is harder to close the operation down, even if its effects are shown to be harmful to wildlife, than not to have had the jobs created in the first place.

The limited and provisional nature of scientific knowledge thus makes it hard to respond satisfactorily to innovative proposals. But 'not knowing' can also be disadvantageous because of the impact it has on the public. Members of the public may look for authoritative judgements and may be dismayed by 'factlessness'. They may look to the scientific experts in conservation organisations for the answers to questions which concern them personally and be frustrated because the 'experts' do not know. Birdwatchers, and even sportsmen, may be concerned about the reasons for fluctuations in bird populations. They see this as the kind of question which conservation scientists should be able to settle, but the scientists may well not know nor even be sure how to find out.

Science may thus be a poor ally to environmentalists because scientists find their lack of knowledge exposed. Sometimes they may lack the ability or expertise to determine something which they wish to know; at other times they may know less than the public would like them to know. Unlike a traditional or charismatic authority they do not have the ability to respond to every contingency.

In some measure, such 'shortcomings' may well be common to all forms of the public use of scientific expertise but conservation scientists face these problems in a particularly concentrated way. For one thing, scientists working in environmental pressure groups often cannot commission or carry out the research that they would ideally like to see performed. According to Cramer, such scientists therefore face 'pragmatic uncertainty' (1987, 50). She argues that they are commonly called on to make recommendations at short notice, often using readily available research material which may be very variable in quality. They are frequently without the time or resources to conduct research of their own. Studies of the environmental effects of minute quantities of contaminants, for example, demand a great deal of time since the substances only build up in the environment very slowly. Scientific equipment and research can also be extremely expensive. Environmental groups simply do not have the financial resources to commit funds to major research projects. Even those organisations with big budgets face many conflicting demands. For example, Greenpeace's campaign boats continually require funding for maintenance and crewing. For its part the RSPB (Europe's largest nature conservation organisation) uses much of its income for the purchase of reserves for the direct protection of birds. To spend money on research is to divert funds from direct environmental action. In many cases, environmental groups are not even able to monitor all the potentially useful scientific information that is published in journals; after all, even university libraries subscribe to only a fraction of the available scientific literature.[5]

Furthermore, when they do turn to the academic literature, environmentalists typically find that research reports are directed to answering the theoretical problems posed by the development of scientific disciplines rather than to meeting their own practical queries. As sociologists of science have long pointed out, most academic scientific research is principally addressed to problems that arise within the matrix of the discipline (see Böhme *et al.* 1976). It is not intended for immediate use by 'customers' and the research objectives of academic ecologists are unlikely to coincide with those of practical conservationists (on this point see Yearley 1995b). This distance between the parties is reflected in the style and content of scientific publications.

In addition, Cramer points out that environmental scientists are handicapped by 'the low level of theoretical development of (ecosystems-) ecology' (1987, 50). She asserts that there is less consensus in ecological science than in many other areas of natural science so that the interpretation of ecological information is especially likely to be disputed. Finally, in Cramer's view, environmentalists face a further form of uncertainty which stems from the fact that they are dealing with large-scale phenomena taking place in open systems. The empirical material that ecological science sets out to describe is inherently more complicated than the phenomena addressed by those other parts of natural science, technology and engineering that operate with closed systems where all the leading variables can be closely monitored and controlled.

There is, however, at least one sense in which Cramer's account understates the extent of uncertainty which conservation scientists face. The pragmatic uncertainties which derive from the complexity of the systems with which conservationists deal and from the sheer bulk of material available to them can be augmented by another consideration. There are some matters which are on the margins of observability. Ironically, the significance of this factor is attested to in a roundabout manner by Nicholson in the course of the development of his argument for the importance of scientific knowledge in the progress of the conservationists' case. He summarises early conflicts between environmentalists and various established groups such as oil shipping enterprises (which were not regulating discharges), pesticide manufacturers (who were not aware or careless of the entry of toxins into the food chain) and farmers (who were aiming to increase productivity at the cost of wildlife and habitat destruction) (Nicholson 1987, 44–53). He documents how each of these cases was fought out through argument, publicity and lobbying, and concludes by saying:

> The record shows that environmental conservation sooner or later succeeds in dealing with offences of a tangible character but recently the trend has been towards much more disturbing *intangible damage* to the environment, either through chemicals let loose to roam through different layers of the atmosphere, or through nuclear radiation. The resulting problems are particularly disturbing because they strike so widely at the roots of global life support systems, because their agents *are technically so difficult to monitor,* especially when they combine spontaneously with one another after release, and because of their ability to become transported rapidly over long distances and at varying altitudes by air currents, *whose movements are imperfectly understood* (1987, 54, emphases added).

Environmentalists face a practical difficulty when they have to decide how to respond to 'intangible damage' and to agents which are almost impossible to monitor.

Here again science is a less than perfect friend to environmentalists. If, for instance, the hole in the ozone layer is less pronounced one year than before, is that attributable to that year's odd climatic conditions or does it throw doubt on the trend which had hitherto been taken as proving depletion? Equally, many predictions suggest that the greenhouse effect will lead to greater variation in the weather so that cooling as well as warming could be seen as evidence for the existence of the effect. Even the interpretation of fluctuations in the numbers of migratory birds is subject to the same sorts of problems. Huge flocks of birds are very difficult to monitor in detail; if the numbers at a particular site fall for one season this may be because that local habitat is declining in quality, or it may be due to any number of other factors: harsh conditions during migration, unusual predation during breeding and so on. Admittedly, these may be extreme cases (although it is important to note that some of the leading global environmental issues number amongst them). But they do serve to make clear one central point: the empirical fallibility of science, most pronounced when phenomena are at the limits of observability, means that social problem claims founded on science must offer hostages to fortune.

Illustrative case study: science, climate change and the greenhouse effect

The general nature of the problem of climate change is well known. It seems obvious that if people continue to add CO_2 (and other greenhouse gases such as methane) to the atmosphere the Earth will become warmer on average and then dangers of flooding and climate disruption await us. But the simplicity of this message belies the complexity of the greenhouse effect. Although we can hardly doubt that the human race discharges billions of tonnes of CO_2 into the atmosphere each year, we are far from certain about the fate of all the gases we discharge or about the relative significance of humans' impact on the atmosphere. While a billion tonnes (also known as a gigatonne) sounds like a huge amount, as the science writer Fred Pearce (1989, 114) reminds us:

> the atmosphere contains about 700 gigatonnes of carbon; the planet's soils, forests and plant life hold a similar amount, with perhaps twice as much in soils as in plants. The sea waters hold some 40,000 gigatonnes, 60 times more than the atmosphere. So far, humans have located a mere 4,000 gigatonnes in coal and oil deposits... But all these reservoirs pale beside the carbon in sedimentary rock at the bottom of oceans. That total is around 70 million gigatonnes.

It is easy to see that if any of these other stocks of carbon vary from year to year, even by just a few percentage points, they would override human influence on the atmosphere.

A scientific understanding of the global distribution of carbon is evidently needed but matters are far from straightforward. For one thing, the facts cited in the last paragraph are far from indisputable. Estimations of the amount of carbon in the plants of a rainforest or in African soils are just that, estimations. So too are the facts about the amount of carbon in the world's oceans. These facts are conjectural in just the way I described earlier in this chapter. And when these facts are put to use, even to quite simple use, some of the difficulties become apparent. It is surprisingly difficult just to do the carbon 'budgets' (for an empirical study of the way in which such uncertainties have been exploited in policy debate in the USA, see McCright and Dunlap 2000, 2003). It is all the more difficult to understand how the world's stores of carbon will respond to the greenhouse effect; will the natural stores of carbon tend to absorb the atmospheric surplus or will global warming lead CO_2 to be released from the reserves and thus exacerbate the problem? It turns out that some very complex and unexpected areas of science have something to say on this matter.

It has commonly been observed that the Earth's temperature is kept reasonably constant by the regulation of CO_2. Plants absorb CO_2 and produce oxygen; most other forms of life consume oxygen. This might lead one to anticipate that the Earth's temperature would have tended to remain stable over the millennia. In comparison to other planets this is probably so. But from the point of view of terrestrial animals the story looks rather different. To them, the Earth's history would seem to have been characterised by large temperature variations, including periods of extreme cold in the ice ages. Studies concerned with the causes of ice ages have suggested mechanisms which may govern or at least influence the globe's temperature control. One factor which is believed to be a basic cause of changes in global climate is beyond any imaginable human control: the Earth's orbit around the sun is not completely consistent. In effect the Earth wobbles on its axis in a variety of ways, slightly changing its orientation to the sun a few times every 100,000 years. During some parts of this slow wobble the poles receive less heat than usual. They would therefore be expected to cool, allowing ice to build up. As the polar ice sheets expand they reflect more of the sun's heat back into space, thus intensifying the cooling. The growth of ice might become self-sustaining, at least until the Earth's position shifted and warming recommenced.

If this was the whole story, ice ages would have little to teach us about the greenhouse effect. However, as Fred Pearce explains (1989, 139–56), the oceans may contribute to the onset and to the ending of ice ages through their influence over atmospheric CO_2. This may occur in either of two ways. First, the oceans may have a direct physical influence on this greenhouse gas. On top of all the daily and seasonal currents in the oceans, driven by the wind and the tides, there is one major pattern of oceanic flow: comparatively warm water flows on the surface of the oceans from the north Pacific, past India and southern Africa and up into the north Atlantic.[6] Between Norway and Greenland, where the warm flow (known in the North Atlantic as the Gulf Stream) meets the cold of the Arctic, the water cools and dives to commence the return flow, past the Antarctic to the Pacific. The amount of water involved is vast as is its effect on the atmosphere: 'In the far north Atlantic, twenty million cubic metres of water descend to the bottom of the ocean every second…The north Atlantic may swallow up several billion tonnes of CO_2 each year in this way' (Fred Pearce 1989, 140–41). During ice ages, when water is locked up in ice sheets and thus withheld from the oceans, the operation of this global current may be affected. To take an exaggeratedly simple example, a halt in its flow during ice ages could lead less CO_2 to be absorbed by the oceans. The gas would then accumulate in the atmosphere, the greenhouse effect would begin to set in and the Earth would gradually warm up, hastening the end of the ice age. By the same token, if present-day global warming comes to heat the oceans, this too might impede the oceanic circulation and thus compound our greenhouse problems.

The second anticipated influence on the greenhouse effect is a biological one. Ice ages would clearly have a decisive impact on marine and terrestrial life. It is easy to imagine how biological and climatic factors could interact. To take again a simple example, if the cold began to kill off the plants and animals in the sea which absorb CO_2 (for photosynthesis and for building up their shells respectively), the greenhouse effect would begin. Then, as the temperatures begin to rise, life would proliferate and the greenhouse gas would be withdrawn from the atmosphere. However, it seems that the interaction is far more complex than this (Fred Pearce 1989, 143–50). At first, the arrival of ice ages was probably accelerated by biological factors. Perhaps falling sea levels laid bare more land which could be eroded; perhaps water was locked up in the form of ice, leaving the continents dry and leading to an increase in the amount of dust. Either way, more nutrients (like iron and nitrogen) reached the seas and biological productivity rose accordingly. This withdrew CO_2 from the air and led to an anti-greenhouse effect, cooling the Earth further. Gradually, though, the nutrient supply from land would begin to become exhausted. To some extent the nutrients can be recycled in the surface layers of the seas but slowly they would be lost to the deep ocean. Marine life would then rapidly diminish and its cooling contribution would cease. This interpretation of events is held by some to hold practical implications for the reaction to today's greenhouse problem. For example, if nutrients radically affect the ocean's uptake of CO_2, it might be possible to treat the ocean with chemicals to promote marine life which should lead to reductions in global warming (Fred Pearce 1989, 151).

These brief comments on recent attempts to understand the workings of the oceans and atmosphere can only give a flavour of the scientific complexities involved. They will, though, serve to bear out the claims I made earlier and indicate how they play out on a far larger (indeed planetary) geographical scale. The greenhouse effect cannot be understood and responded to without scientific expertise. But a scientifically authoritative interpretation of the greenhouse effect is still far from us; in some respects, we even seem to become less certain the more we study it. Knowledge about the oceans is hard to come by. We cannot observe the deep ocean current in any straightforward sense. In addition to problems of observation, the 'science of the greenhouse' is hindered by theoretical uncertainty and by the problems posed by an open environment in just the way Cramer anticipated. The climate and the oceans are amongst the most open and complicated systems which scientists have tried to model.

Before concluding this section, I should make clear that these observations are intended not as criticisms of the way that scientists study the climate system nor as a denial of the importance of global warming. I know of virtually no one who argues that we should continue blithely to discharge CO_2 from our motor vehicles and power stations. For one thing, by using our fuel resources more sparingly we reduce all manner of pollution aside from CO_2, as well as lowering energy costs. But my interest in this chapter is in the role played by scientific authority. Science is needed to offer authoritative advice on our environmental problems but for understandable, yet persistent reasons that authority is not as decisive as environmentalists might desire. O'Neill has taken me to task for presenting science as an inadequate 'friend' to the environment, claiming that no friend could be so reliable as to never let one down (1993, 146). But the point is that the practical weaknesses of scientific authority commonly manifest themselves most forcefully at just those points where environmentalists would most like to be able to rely on science to deliver 'the facts'. For this reason, environmentalists find themselves in a situation surprisingly similar to that of other groups making social problem claims. They invoke scientific considerations and try to enlist scientific authority but cannot look to science for the decisive proofs they would dearly like to have.

Science as an unreliable friend – epistemology

Up to this point we have examined ways in which *in fact* science may be a less good friend than conservationists might anticipate. It may not provide the answers on occasions when it would be politic to have them; it may leave the public impatient of factlessness; and some of the facts that conservationists might like to marshal may be elusive. But in some cases these deficiencies come close to endemic problems of scientific knowledge to do with science as a way of knowing at all. Most nature conservationists would defend science as a form of knowledge by pointing to its observational basis and its methodic development. But, as we have seen and as Nicholson made clear, the observational basis is open to discrepant interpretations. As soon as there arise competing and plausible accounts of what the observational facts are, the basis which appears so secure becomes itself problematic. The empirical and provisional basis of scientific knowledge – its apparent strength – can readily be reformulated as an uncertain basis. This argument achieves its most spectacular form when it is posed as the philosophical problem of 'induction'. For centuries philosophers have pointed out that although, for example, we may believe that the sun will rise tomorrow because it has risen every day so far, this can only be an assumption. We cannot know such things for certain. In the past, traditionalists tended to use such arguments to contrast science unfavourably with other paradigms of knowledge, like religion or logic. Logical deduction appears more certain than empirical induction.

In recent years these arguments have not been regarded as having much practical importance. People have not worried about the likelihood of there being a sunrise tomorrow. But this line of thought does show up the Achilles' heel of science, a weakness which can be used in a practical way to evade scientific authority. Those opposed to a scientific judgement can always say that science is not fully certain and that, for this reason, they do not recognise expert scientific opinion as ultimately authoritative.

As one might expect, environmentalists tend to play down this issue in many public contexts. In a general publication such as the *BBC Wildlife Magazine*, Jonathon Porritt (a prominent UK green and former director of Friends of the Earth (FoE) wrote:

> the scientists are now with us rather than against us. On occasions...they actually seem to be out in front of the activists of the Environment Movement. In the early seventies, the protagonists of the 'limits to growth' scenario relied primarily on an inadequately programmed computer model. Politicians had little difficulty dismissing it as sensationalist speculation. Today, there's nothing speculative about the depletion of the ozone layer, the deforestation of the Amazon, the build up of CO_2 in the atmosphere, or the pesticide residues in our water and food. Hard scientific evidence counts for a lot in a hard materialistic world. (1989, 353)

It is interesting to note that scientific proof is associated with hard materialism, almost as though scientific support for environmental policies would not be needed in a more compassionate society. None the less, this embrace of the 'hardness' of scientific evidence displays the kind of strategy which can be adopted in chasing off the horrors of doubt.

Other strategies may be adopted. In the same year Greenpeace ran a newspaper advertisement campaign in the UK opposing claims by Mrs Thatcher's Environment Minister, Nicholas Ridley, to the effect that increased investment in nuclear power generation would help solve the greenhouse problem. The minister was pictured with his pro-nuclear assertion printed across his mouth; beneath it was written: 'scientifically speaking, it's just a lot of hot air'.[7] Greenpeace then printed a declaration disagreeing with Mr Ridley, a declaration apparently signed by '100 of the country's leading scientists, doctors, and engineers'. Now, there is something curious about the logic of this move for although Greenpeace are invoking scientific authority, there is a majoritarian appeal also. Their argument seems to be not just that 'scientific opinion' is with them, but that *a lot* of scientists think this way. Yet, one could just as easily argue that, in the context of the UK, one hundred 'scientists, doctors and engineers' is actually very few. It seems that Greenpeace is forced into this roundabout appeal because they are confronted by an interpretative difficulty. They wish to claim to be in the right epistemologically – to say that Mr Ridley is simply wrong. Yet, while the epistemological right may in principle be straightforward and unambiguous, in practice both sides in any dispute can usually count on some scientific supporters. Both sides may try to claim the epistemological high ground. An appeal to large numbers of qualified supporters is perhaps the simplest way to respond to this difficulty in a mass public medium.

That this way of trying to overcome the problem of identifying the scientific high ground is not unique to Greenpeace is indicated by a subsequent experience of FoE. Campaign staff working on climate change issues were disturbed by a programme aired on the UK's Channel 4 in the 'Equinox' series in 1990 that sought to question the scientific evidence for greenhouse warming. The programme even implied that scientists might be attracted to make extreme and sensational claims about the urgency of the problem in order to maximise their chances of receiving research funding. The programme was criticised in the 'campaign news' section of the FoE magazine, *Earth Matters*. An unfavourable comparison was drawn between the sceptical views expressed in the programme and the conclusion of the Intergovermental Panel on Climate Change (IPCC), which had warned of the reality of impending climate changes and with whose scientific analysis FoE was generally in agreement. FoE's article invoked the weight of 'over 300 scientists [who] prepared the IPCC's Science Report compared to about a dozen who were interviewed for Equinox'.[8] When apparently well credentialed scientists are seen to disagree it is very difficult to claim simply that one is in the right. It seems like a reasonable alternative to invoke the power of the majority. But of course this cannot always be done since, as the quote from Porritt made clear, in many areas where environmentalists believed themselves to be factually correct, they had been in the scientific minority, at least initially.

Whatever the complexities of enlisting scientific authority in a positive fashion, from the conservationists' point of view the difficulty is most acutely felt when it goes the other way – when they are confronted with the barrier of scientific proof. As Richard North, science and environment writer for *The Independent*, has noted: 'Even now, scientific exactitude can debilitate conservation by insisting (as governments rejoice to notice) that the evidence of damage caused must be total: which it almost never will be' (1987, 15). This argumentative strategy is probably best known in the UK through its use in relation to acid rain. Throughout the 1970s and much of the 1980s, the authorities used the lack of certain knowledge that acid rain (and in particular British acid rain) was responsible for the death of trees and the acidification of lakes in Europe as a justification for continuing with power-station emissions (Yearley 1992a, 107–09).

Different groups adopt different responses to this problem. Greenpeace, whom North was criticising in his article in *The Independent*, tends to be impatient of the limitations of scientific proof. North accepts that Greenpeace may be correct to suppose that it 'would get little [media] coverage were it to stick to the facts. In any case, it does not' (1987, 15). He goes on to list ways in which Greenpeace's public statements have been, as he puts it, 'economical with the truth'. His contention seems to be that Greenpeace has often bent the scientific truth to make issues appear graver than they truly are in order to stir people into action (see Chapters 9 and 10). This bending has, for example, taken the form of over-generalising from atypical examples. It remains unclear what strategy North would endorse, given the uncertainties which inevitably accompany 'good' scientific practice (for reflection on this from Greenpeace's point of view see Rose 1993). Moreover, it should be appreciated that activist organisations can point to a number of occasions on which their gloomy expectations have been vindicated when further spills or leaks have followed assurances that the 'problem' has been overcome. With some plausibility they could argue that there is a practical asymmetry between doing nothing (the situation will worsen) and doing something (it may bring unnecessary cost but no permanent damage to the environment). Other organisations adopt a self-consciously contrasting tone. For example, during my fieldwork the chairman of the UWT characterised his group as standing for 'informed, educated, reasonable, rational conservation'. This seems to advocate a different epistemological stance from that adopted by Greenpeace but the problem, of course, is how to confront urgent but uncertain issues rationally and reasonably.

Within the last decade, the idea of using the 'precautionary principle' has risen to prominence as a supposed solution to this problem (for a study in the European context see Dratwa 2002). The idea of this principle is that one should switch the balance of proof onto those who propose to take the action which has the possible environmentally harmful side effects. As will be seen in the next two chapters, this principle has come to be widely invoked, for example with regard to the deliberate release of genetically modified (GM) crops. On a precautionary approach, a company or a government that wishes to do something novel must show that it is not harmful to the environment before permission would be given for it to go ahead. This is radically different from the liberal assumptions of current planning procedures which permit anything that is not specifically illegal, and then tend to assess harms once they have become apparent.

However, environmental groups that espouse this principle commonly play down the ambiguities associated with it. For one thing, it is unclear how a company could prove that some new process or product will do *no* harm. There is firstly a logical problem in proving a negative result. How could one show that no harm would ever result? More practically, if the harm is very subtle or slow to appear (as it was with CFCs), one might need decades of testing before one could be reasonably sure that no harm will result. Furthermore, many innovations are associated with both harms and benefits. Innovators tend to complain that the precautionary principle pays more attention to possible harms than to likely benefits. Accordingly, even if one adopts the precautionary approach, a great deal of variation can arise in the way that the principle is interpreted. Advocates of the principle often speak as though the principle were self-evident. It is not; it requires every bit as much interpretative work as other approaches.

The specific epistemological character of science thus leads to difficulties when groups try to 'cash in' on scientific authority. If you are relentlessly committed to scientific proprieties, you will not be able to make instant, unequivocal judgements. But if you are not publicly committed in this way, you are open to criticism. The resulting practical difficulties may become apparent during scientific disputes or in public controversies. But they can also surface in a virulent form in specialised forums of debate. In particular, the conventions of legal cross-examination and the standards of legal proof may not mesh well with the character of scientific argument and expertise. The way this factor works out in the British case was fully illustrated in the case studies in Chapters 5 and 6, where barristers acting for developers successfully combated the scientific evidence of nature conservationists (Yearley 1989, 1992b). The role of scientific expertise in legal disputes relating to the environment has often taken a different form in the USA as the next section will show.

The US experience: a contrasting culture of proving

In the USA, not only were regulatory bodies such as the Environmental Protection Agency (EPA) established relatively early (at the start of the 1970s), but they took steps which were seen as radical at the time, including pressing for the adoption of scrubbers in coal-fired power-station chimneys and for the fitting of catalytic converters to car exhausts. Before long, these extensive reforms stimulated rearguard action on the part of industry, which argued that suspect claims about environmental damage were being allowed to justify the introduction of commercially damaging regulation. This theme won political support at the highest levels during the Republican administrations of presidents Reagan and Bush in the 1980s and again with the Republican Congressional majority in the 1990s.

Aside from political manoeuvring, both industry and environmental groups pursued their regulatory interests through the courts, marshalling counter-expertise to combat the judgements and technical assessments adopted by the EPA and other bodies. The availability of citizen suits and other judicial remedies meant that pressure groups in particular found themselves in a very different context from that prevailing in Britain. The obvious role for them to adopt was as a prod to the EPA, and sometimes as an explicit counterweight to industry interests.

Given the resources which industry could devote to challenging environmental regulations and the high stakes involved in these confrontations – for example, a ruling that formaldehyde is carcinogenic to humans would have affected a billion-dollar industry in the 1980s (Jasanoff 1990, 195) – it is no surprise that disputes over scientific evidence were fought tenaciously and with great inventiveness. Since the competing sides could find either plausible-enough experts to support any position or, at worst, authorities to throw doubt on their opponents' views, these trials demonstrated only too well that scientific standards of proof do not in practice enjoy the authority one might expect of them in principle. Since these challenges were channelled through the courts, technical disputes over health, safety and environmental hazards were all opened to judicial – and hence publicly documented – scrutiny.

These challenges to the expert views of the EPA and associated agencies were noteworthy not just for their impact on the development of environmental politics but for what they revealed about the strengths and weaknesses of scientific reasoning. Sociological and philosophical analysts of science have, for the last two decades, been analysing scientific controversies to understand how accepted scientific beliefs come to face opposition, deconstruction and overthrow. Exactly analogous processes were revealed by these legal challenges. New proposed tests for toxicity faced deconstructive challenges, as with innovative experimental tests for any physical phenomenon (Collins 1992, 2). The very same difficulty is faced by agencies attempting, for example, to carry out toxicity assessments; this is because, until some test has won acceptance, there is no separate touchstone of credibility. Courts may be wary about accepting the evidence of innovative tests because the tests themselves have not won widespread acceptance. Yet in the absence of the tests, harmful substances may go unregulated (see Jasanoff 1995). This is a logical impasse. This problem is bad enough in 'pure' science, where the reasons for distrusting others' results are disciplinary or occasionally personal. The disagreement may turn into an acrimonious controversy but the scientific world can wait for the answer as the persuasive resources of the competing sides are marshalled and developed. In disputes over environmental safety, there is typically considerable urgency about resolving the issue; at the same time, huge commercial and political motivations may also be involved, creating further incentives for discrediting the opposing side's claims to scientific knowledge.

When agencies, faced with repeated and protracted legal opposition, ran into serious problems with their public credibility it was common for a review to be instituted, the typical conclusion of which was that the agency was conflating issues of science and policy. The approved remedy was to take various administrative steps to segregate these activities. The EPA's own Science Advisory Board (put on a statutory footing in 1978 [Jasanoff 1990, 84]) became a key element in meeting these demands for segregation. However, as Jasanoff has convincingly argued, such segregation cannot be achieved because there is no 'true' boundary to be found. For example, evidence that substances are risky to humans comes – in large part at least – from animal toxicity studies. For each substance and each combination of substances, it is just possible that some aspect of rats' biology (their nasal tissues, their kidneys or some other system) differs from that of humans. Treating animals as models for the human impact of potential toxic substances is thus a practice based on a reasonable precedent, but it cannot be relied on for any particular case. Where it works in their favour, industries are inclined to accept the validity of the test. When it does not they are inclined to query it. The validity of using data about toxicity or carcinogenicity in rats as compelling evidence in relation to humans is thus both a matter of policy and of methodology. The conundrum of how to separate (compelling) scientific arguments from (vulnerable) policy ones cannot be resolved simply by people trying to be 'more scientific'.

There are of course some peculiar features to the scientific issues which the EPA and other environmental agencies often have to regulate. They deal with quantities that are hard to measure, physical phenomena that are highly interactive, and diseases that occur over the course of a lifetime and for which there may be many plausible causes. The science involved in such determinations lends itself to controversy (see Collingridge and Reeve 1986). But the point revealed in a series of legal challenges is the disputability of scientific knowledge per se, not the special disputability of the science of cancer or of pesticide toxicity.

It is only in regulatory systems which are characterised by secrecy and confidentiality that the appearance of scientific authority's impregnability is maintained. In her exhaustive studies of such conflicts in the USA, Jasanoff reveals that benign decision-making often takes place when institutional arrangements offer a temporary respite from the endless, adversarial legal review. The European experience of course is that official agencies often are not trusted and many environmental groups hold US standards of freedom of information up as a model. Overall, therefore, the paradoxical result is that in the one area where scientific expertise has been most often used for environmental-policy purposes, neither secrecy nor competitive openness have proven to be suitable climates for giving authoritative advice to policy makers.

Science as an insufficient friend

The ecological movement's dependence on science may not bring it the cognitive authority it seeks either in practice or principle. But there are some environmental arguments where its insufficiency seems to be of another sort. Take the case of the Marine Conservation Society's (MCS) continuing concern for the basking shark. The MCS says that this shark is an animal which does no harm to humans and which is majestic in its own way. Sadly, UK (and European) fishing practices endanger it and we should protect it for its own sake. At present, relatively little is known about these creatures so that the lack of empirical information is a handicap to the campaign. But, let us suppose that we did know all about these sharks and that they are actually in danger. It might still be possible to argue that the sharks are not worth the loss of fishing. What this suggests is that the arguments for the environmental case are far from solely scientific; indeed they look rather similar to those surrounding abortion and artistic freedom.

To put this another way, the scientific correctness of the environmentalists' analysis – even when that correctness is undisputed – still does not carry many clear implications for practical action. If it is believed that the destruction of the ozone layer will result in many human deaths then possibly there is a common practical need to take steps to remedy the problem. But in the case of the majority of other environmental problems the practical implications are more disputable. There may be many reasons for trying to conserve the rain forests: for the sake of the tribal people who live there, for the sake of the plants and animals themselves, on account of the likely medicinal value of rainforest plants and so on. The same variety is true for species conservation: elephants can be valued in their own right as well as for the possible economic benefits of the tourism associated with managed herds.

When mounting campaigns, environmentalists tend to invoke as many of these reasons as they can in order to attract the greatest number of supporters (Yearley 1993). As noted in Chapter 2, people may join a group or give money for a wide variety of reasons. The RSPB appeals to naturalists, to bird enthusiasts and to those with a diffuse concern for the countryside. Greenpeace attracts animal lovers as well those opposed to the nuclear industry. Although, therefore, science is used in making social problem claims, science itself does not dictate which claims will be made. For instance, the RSPB draws on considerable scientific skills in conserving birds but it was not scientific reasoning which led the group to work for birds rather than field mice and voles. Science is not a sufficient guide to what conservation groups should concentrate on and prioritise; nor, often, does science provide the members' reasons for engaging in conservation activities. In a narrow sense, science does not seem to compel people to conserve particular bits of their environment nor tell them what the conservation priorities are.

Remoralising science: Gaia

Up to this point I have suggested that science seems not to offer a moral basis for the green case. However, one popular idea manages to tie the science of the environment to moral concern for the planet, the Gaia hypothesis. This hypothesis can best be explained through an example. Commentators describing the greenhouse effect commonly point out that the CO_2 in the atmosphere seems to have regulated the Earth's temperature. Without the greenhouse effect, the Earth would probably be too cold to support life; it would resemble Mars. That many life forms have persisted for hundreds of millions of years implies that the Earth's temperature has long been regulated. To talk in this way about 'regulation' might seem suspect and anthropomorphic. After all, there was no one there to do the regulating. However, many commentators are content to use this vocabulary, regarding the description only as a metaphor.

But the Gaia hypothesis proposes that it is not just a metaphor; according to this hypothesis the Earth really is regulated. Life on the planet is somehow co-ordinated in a way which works to keep the planet habitable. On this view, the planet should properly be regarded as a super-organism. Jim Lovelock who proposed this idea gave the name Gaia (the Earth goddess) to the workings of this super-organism. Clearly, if the Earth can be described in this way, it might change our expectations of the globe's response to human meddling and alter our attitude to the planet; we might now see ourselves as having a moral obligation not just to humans and animals but to Gaia.

Scientific supporters of the Gaian idea would claim that it has helped them understand the ecology of the Earth through its emphasis on the contribution which living organisms make to the maintenance of the planet. Gaian ideas stress the extent to which the whole planet is suffused with life. Surprisingly few physical processes operate without the intervention of living organisms. For example, we have already seen how organisms in the sea may play a central role in determining the Earth's temperature by controlling the amount of CO_2 in the atmosphere. Similarly, soils are not simple, inorganic materials. Organisms that live in soils are essential to their biological productivity. Thus, a demythologised Gaia hypothesis might amount to a belief amongst scientists that living things are empirically more important to the physics and chemistry of the Earth than has hitherto been recognised.

However, this demythologised version is not the only one in circulation. More overtly holistic Gaian ideas have proved attractive to some scientists and to many in the environmental movement (see Porritt and Winner 1988, 249–53). At the same time, other scientists have perceived these ideas as deeply disturbing. Sceptical scientists raise both empirical and logical doubts. At an empirical level, for example, one might seek to query the hypothesis by citing the contribution which organisms appear to have made to the intensification of global cooling at the start of ice ages. Surely, it might be said, Gaia would counteract rather than assist such cooling. But most arguments have been addressed to the logic of the Gaia hypothesis. Thus, it is argued that there are problems about the idea of ascribing purposes to Gaia. If it is to be more than a metaphor, Gaia would seem to need to have purposes in the way that only humans and (conceivably) a few animals do. Scientists do not treat plants or bacteria as though they have deliberate intentions; how then can the planet possess them? Even, sceptics argue, if one accepts that it might be possible to talk of the planet's purpose, it is impossible to know what that purpose is. We cannot communicate with the planet. As Horsfall asks, 'what is her purpose to perpetuate the rock endoskeleton, to ensure the safety of the majority of species, to ensure the safety of certain key elements such as bacteria, or to ensure the safety of [humankind]?' (1990, 25). From time to time, as we have seen, there are ice ages and other catastrophes which result in mass fatalities. How can these be reconciled with Gaia's assumed purposes? Of course, one could say that the purpose is not to retain any particular life form but an aggregate life force but, since we do not know how to count this aggregate, we could never test such a theory.

Many scientists felt indignant about Lovelock's claims; for several years his papers were routinely turned down by leading journals though he has won a few prestigious scientific supporters (Fred Pearce 1989, 37). Still the pragmatic strength of the Gaia hypothesis is that it offers to combine environmental science with morality. It thus seems to provide the kind of authority which, as we have seen, a more routine dependence on science fails to deliver.

Conclusion

To regard the environmental movement as profoundly anchored in science is surely correct. But, in practical terms, ecological campaigners have found it far harder to cash in on that scientific authority than might have been anticipated. In this chapter, I have sought to explain why science is less of a cognitive ally to greens than they might hope.

In part, the explanation is philosophical. Scientific knowledge is inherently open to revision; it is intrinsically provisional. Particularly at the forefront of science, it is always possible that the truth is at odds with scientists' current beliefs. Despite science's cognitive power, it cannot offer transcendental support for particular substantive propositions. Moreover, the environmental movement is dependent on extra-scientific, moral considerations. Scientific studies indicating that whale populations are declining to non-sustainable levels may well offer good grounds for not whaling. But when (as has recently happened) populations begin to recover, scientific reasoning does not suffice to say whether or not hunting should be resumed.

The explanation is also sociological. The profession and the practice of science mean that the research that greens desire or need may not be done. The social composition of environmental groups may not afford them the scientific expertise they require. Governments, firms, unions – even campaigners – may be far from disinterested in the uses they make of scientific information. Finally, philosophical and sociological factors may overlap and interact. The social context of legal inquiries encourages the tendentious exploitation of science's epistemological weaknesses; media conventions about 'fairness' encourage broadcasters to give 'equal time' to competing views even if the scientific credentials of those views are far from equal.

Together these sociological and philosophical factors help explain why even a social movement with profound scientific support experiences difficulty in winning over the authorities and bringing about policy changes. In turn, this experience of frustration – when science fails to deliver the expected benefits – is likely to reinforce environmentalists' attitude of ideological ambivalence towards science and to stimulate demands for alternative sources of legitimation. The tension between a scientific professionalisation of movement organisations and the pursuit of more comprehensive political and philosophical (even spiritual) legitimations is set to continue.

9
Mad about the Buoy: Trust and Method in the Brent Spar Controversy

Introduction: witnessing, observation and objectivity in science[1]

Philosophers assure us that the aim of scientific investigation is to arrive at an impartial and objective understanding of the nature of the world. Of course, all manner of practical and intellectual obstacles can get in the way of this aim; such obstacles may even get in the way more often than not. But the contemporary conception of science is inseparable from the idea that it sets out to describe how the world is. Disciplined observation of the way things are in nature is obviously a key element in the pursuit of scientific understanding. Of course, science is not only about observation and experiment; it is also about theoretical analysis and conjecture for example. But it could hardly thrive without observation.

However one of the achievements of recent work in the history and sociology of science has been to reveal how complex the business of observation is. Indeed, at the time of the 'scientific revolution' in the seventeenth century the idea that experimental observation was the best way to arrive at an understanding of nature was rather disreputable, in most respects a minority opinion (see Shapin 1996, 109). In some circles this was because the study of learned texts was highly esteemed. Among other thinkers this scepticism arose because it was assumed that the exercise of reason alone could produce large gains in understanding. But it was also because it was so obvious that observation for scientific purposes was fraught with problems. Observations were prone to many sorts of error. If they depended on specialised equipment, the equipment could itself introduce unknown forms of distortion. Worse, observations were private. Champions of the empiricist approach to knowledge of nature tended to favour an image of individualism in which the observer is convinced by what he (or, rarely, she) sees: '[Theirs] was a rhetoric which insisted that no source of factual information possessed greater reliability or inspired greater confidence than the direct experience of an individual' (Shapin 1994, 202). Yet this individualism clearly held its own dangers. If I saw something in my study or laboratory then that did not mean it was available for others to see. By contrast, if I deduced something through reason alone then others could arrive at the same insight by doing the reasoning themselves. Troublingly for empiricists, some observers could be sincere but unreliable; since all observation is private, there was a persistent problem in knowing whose testimony to accept and whose to reject.

In their pioneering work on *Leviathan and the Air-Pump* Shapin and Schaffer (1985) investigated the role of a new form of experimental device, the air pump, introduced in the late 1650s, which facilitated experimental observation in the ensuing decades (see also Shapin 1994). Previous historians had concentrated primarily on the details of the air pump itself. It was an enormously difficult thing in the seventeenth century to make a large glass vessel with the necessary characteristics to serve as an experimental air pump. The glass container and its seals had to be air-tight enough to allow the great majority of air to be pumped out successfully. Yet the inside of the vessel had to be accessible to permit one to be able to place things in the apparatus in order to conduct experiments on them – for example to see how long a candle would continue to burn when air was excluded or how a barometer behaved in a near-vacuum. These demands for ease of access and air-tight construction threatened to contradict each other and to invalidate many experimental observations. But, as well as noting these technical matters, Shapin and Schaffer directed their attention to the way in which early experimenters sought to overcome the problem of how to turn their observations on an admittedly fallible machine into knowledge that would be robust. Besides the material technology of the air pump, experimenters of the second half of the seventeenth century needed to develop a 'social technology' to secure the role of observational evidence.

Briefly expressed, and drawing also on subsequent work by Shapin alone (1994), their argument is that this observation evidence was made robust in three ways. First, social conventions about the good 'word' of the English Gentleman were used to turn gentlemen's accounts of what they had observed into more than just casual reports. A gentleman's word was not to be doubted. Gentlemen commanded trust, and with a little work that trust could be extended to their observational reports.

The fact that a gentleman observer might be accompanied while making observations by his employees, servants and even women-folk contributed little to the robustness of his testimony. A servant would – of course – insist that he saw what his master claimed to see; that was the servant's obligation. But the force of the observation could be increased by having numerous gentlemen observe it. This was the second social technology: occasions for experimental demonstrations were set up so that guests of elevated social standing could join in the observation as well. This was institutionalised in the experimental sessions of the Royal Society of London, where 'a *Register-Book* was provided for witnesses to testify their assent to experimental results' (Shapin 1996, 107). However, neither at experimenters' homes nor at the Royal Society was it common to be able to assemble very large numbers of gentlemanly observers. Accordingly, as a third form of social technology, a style of writing observational reports was developed, a style that stressed the actualité of the observations. The reader of such reports became a 'virtual witness' to the same phenomena.

> Virtual witnessing involved producing in a reader's mind such an image of an experimental scene as obviated the necessity for either its direct witness or its replication. In Boyle's experimental writing this meant a highly *circumstantial* style, often specifying in excruciating detail when, how, and where experiments were done; who was present; how many times they were reiterated; and with exactly what results... Such a prolix style might 'keep the reader from distrusting' the outcomes related and might assure the reader of the specific historical reality of factual particulars. (Shapin 1996, 108; original emphasis; see also 1984)

In these ways the precarious status of observations was gradually transformed so that observational reality became paramount in the certifying of knowledge. In the modern period, observation tends to be the pre-eminent form of evidence. Nothing is as sure as the 'plain fact' that which can be seen; tradition and contemplation tend to come a poor second to observation as a way of guaranteeing that something is, in fact, the case.

Contemporary cultural attitudes and practices so enshrine this respect for observation that is generally necessary to revisit the state of affairs in the seventeenth century to see how problematic observation could be. Without the presumption of a trusted class of report-makers, without the opportunity for many witnesses to participate in observation and without effective and widely accepted techniques for virtual witnessing, the 'ancient' anxieties about the privacy, unreliability and inscrutability of observational claims could reassert themselves. From these rather unexpected beginnings, and using the notion of these 'social technologies', I hope to throw fresh light on the celebrated case of the controversy over the Brent Spar oil platform.

The Brent Spar

The case of the Brent Spar was perhaps the most publicised single environmental controversy of the last decade, certainly in Europe. The Brent Spar itself was an oil-storage buoy of huge proportions, which was moored in the North Sea between Britain and Norway from the mid-1970s; its function was to hold oil as it was extracted from the Brent field and to allow that oil to be pumped into the tankers that would transport the oil ashore (see Bennie 1998). As the oil industry developed, a network of pipelines was established which allowed the oil to be pumped from wells directly to land-based facilities; the buoy thus became increasingly redundant. The Brent Spar was the size of a small ship, over 100-metres long and 29 metres in diameter. It sat upright in the water like an enormous fishing float, with thousands of tonnes of ballast in the bottom to keep it vertical; it was tethered to the sea floor to stop it drifting away. Approximately 80 per cent of it lay beneath the sea's surface and this submarine component was largely given over to oil storage.

By the beginning of the 1990s it was becoming clear that Shell, the company responsible for managing the buoy (though it was co-owned with Esso), would have to decide on the Brent Spar's future. If it was to stay at sea it would need to be overhauled and re-licensed; if it was to be retired the company would have to determine how to decommission it. According to European statutory requirements, plans to dispose of large structures such as the Brent Spar must be agreed by the appropriate ministry in the relevant member country; companies are not free just to decide for themselves. Thus Shell had to undertake a review of possible disposal methods. The pros and cons of these methods were reviewed by a series of engineering and environmental consultants. British procedure demanded that Shell figure out what is known as the Best Practicable Environmental Option (BPEO), that is the 'best practicable environmental option' – loosely expressed, the best option for the environment at reasonable cost. After their review, the company identified the BPEO in 1994 and this was then ratified by the relevant ministry, the UK's Department of Trade and Industry (the DTI). The option found to be the 'best practicable' one was deep-sea disposal. The buoy would be emptied as far as possible and then towed out of the North Sea, across the top of Scotland and out into the North Atlantic where it would be filled with water, holed by explosive charges and slowly sunk onto the sea bed – still within British waters – in a deep trough. There it would gradually disintegrate causing, it was reckoned, limited local pollution.

This was deemed to be the best practicable option in environmental terms because there was concern that the obvious alternative, bringing the buoy onshore, would pose problems of two main sorts. There would first of all be potential difficulties in turning the structure on its side, something that would be necessary as it could not be towed into port upright; no docks would be sufficiently deep. But rotating the structure to make it horizontal would place enormous strains on the metal hull and the buoy might fracture and cause pollution close to the surface. According to Shell's engineers and advisers, such pollution might well be more disruptive than that anticipated on the deep sea floor in the course of ocean disposal. Attempting to lift it higher in the water as an alternative to tilting it onto its side was also thought to risk accidental releases. Second, even once the buoy was successfully turned onto its side or raised in the water, there would still be problems caused by bringing it ashore. Workers engaged in dismantling it would be exposed to possible hazards, including some slightly radioactive substances (as described below), and the residual material, oil and bituminous deposits, would have to be disposed of somewhere. These and related considerations appeared to have satisfied the British officials about the wisdom of the disposal method selected. And since there were very few buoys of this sort in the North Sea, with no more being planned as they had been rendered obsolete by the pipeline network, the argument was offered that this policy set no significant precedent. Oil drilling rigs were of a different construction and their disposal would be governed by different considerations. Other European governments that were parties to the Oslo Paris ('OSPAR') Convention governing sea disposal in the North East Atlantic were notified and, as no official objections were received, plans were made in 1995 to go ahead with the disposal on the sea bed.

Although some conservation bodies, along with fisheries interests, had been consulted in 1994 as the disposal plans were being developed, Greenpeace apparently only learned of the plans in early 1995 and began right away to orchestrate their opposition (Rose 1998, 44).[2] They objected to the proposed decommissioning method because it represented the continuation – albeit in a new form – of the practice of disposing of industrial waste at sea, something against which they had protested for many years. They were also concerned that allowing one oil-industry installation to be dumped on the Scottish continental shelf would condemn that area to become the graveyard for North-Sea petroleum industry waste. At the end of April 1995, after the British government had announced its intention to permit sea disposal but before the actual issuing of the licence entitling Shell to begin operations, Greenpeace campaigners boarded the buoy. They realised it was better to start any campaign of opposition before the towing was underway. Despite court orders, the Greenpeace occupiers declined to quit the Brent Spar; in the end they were physically evicted nearly a month later. By this time, the German government had officially complained to the UK government about the disposal method adopted (Thomas and Cook 1998, 7) and Greenpeace personnel on board the buoy had carried out their own reassessment of the amount of noxious material still remaining in and on the structure.

Once the protestors were evicted the campaign turned to a different strategy. Greenpeace encouraged a consumer boycott of Shell petrol stations. Of course this was not too demanding for motorists; they did not have to forgo petrol, all they had to do was to buy fuel from an alternative petrol station. Moreover, because Shell is a transnational company with Shell garages across Europe, the boycott could operate also in Germany (in this case apparently in line with the national government's own [recent] misgivings) and in several other countries. Still Shell persisted with their disposal plan, fully supported by the British government, and the tow was begun in early June. But the following week Greenpeace announced the result of its tests on the material on the buoy: they claimed that the Brent Spar contained thousands of tonnes of dirty oil. This raised the stakes hugely since it implied that Shell, with the UK government's connivance, was going to engage in grossly polluting behaviour and – moreover – intended to do so covertly.

Method and measurement on the Brent Spar: social technologies at work

A key element in the arguments presented by Shell and their advisers (notably the consultancy group Aberdeen University Research and Industrial Services Ltd – AURIS), and subsequently emphasised by the British government, was that Brent Spar was essentially empty. Of course, the buoy contained a huge amount of steel and some other metals; indeed it was so big that just the paint on its exterior surfaces contained 3.5 tonnes of zinc. There were estimated to be tonnes of copper wiring and even a few kilograms of cadmium and nickel. But the oil had all been pumped out and flushed with sea water that itself had been taken away for safe disposal. It was now mostly full of more sea water, contaminated to only a very low level with the last residues of oil. There were also tonnes of ballast material mixed with sand and oily sludge. But Shell maintained they had removed every last thing they safely could; they had even taken out the light bulbs from the offices and accommodation areas. Finally, 'the internal pipework in Brent Spar contains LSA scale [very mildly radioactive scale], which is formed when naturally occurring radioactive material dissolved in reservoir water precipitates on the inside of pipes' (AURIS 1994, 1). In other words, the buoy had small amounts of radioactive material in it but this radioactivity had not been produced on purpose nor was it directly related to the core activities of the petroleum industry. It was left as a residue when sea water passed through the pipes. Of course, accidentally present or not, this material would still be potentially hazardous to anyone dismantling the Brent Spar or to creatures that came into prolonged contact with the residue of the buoy wherever it was disposed of.

In mid-June 1995 Greenpeace announced that, during their lengthy stay on the buoy, they had been able to make a reassessment of the Brent Spar's actual contents. In particular, they had sampled a shaft running down, as they supposed, into the oil storage tanks and found that their observations showed that there was vastly more oil on board than Shell had indicated. Greenpeace reckoned that the buoy contained some 5000 tonnes of waste oil. This was very significant since it suggested both that Shell was willing to dissemble about the contents of the Brent Spar in order to dump the buoy at low cost *and* that the sea dumping of the buoy would likely cause a good deal of pollution, not the very small amounts that Shell had anticipated.

Shell's representatives were dismayed about this claim since it reflected so badly on their truthfulness. The situation can helpfully be understood in relation to the three 'social technologies' outlined earlier. First, at the time of the measurement, Greenpeace had effectively been the sole occupants of the buoy and thus had a monopoly on opportunities for observation. But, just as at Boyle's laboratory, the number of observers present was very limited. Greenpeace kept only a handful of people on the structure. They were in a rather better situation in regard to the second component of the social technology: their standing as trusted observers. Unlike Shell, the Greenpeace team seemed to have no vested economic interest in minimising estimates of the amount of oil on board. Their whole campaign of boarding and occupying the buoy, hanging on despite determined assaults with high-pressure hoses and so on, echoed earlier struggles on the high seas with whaling fleets and with ships intent on dumping industrial waste into the sea (Yearley 1992a). In the past they had managed to win public trust through their characteristic combination of derring-do and environmental advocacy; this case looked no different. They appeared to be highly reputable agents, putting forward a simple and direct observational claim, just as they had in the past, namely that the buoy was grossly contaminated with oil.

Thirdly, they could mobilise techniques of virtual witnessing. These were very far from those pioneered by Boyle but performed a similar function. Through their skilful use of media images – they were supplying news companies internationally with filmed footage aboard the buoy – there were millions of virtual witnesses for their observational reports. They could describe what they had seen, relate how they had come to see it and display the telltale signs of the oil residue. As Rose points out, this monopolisation of the news images was not so much engineered by Greenpeace as produced by caution, or under-estimation of the importance of the story, by television news chiefs. There were simply no film journalists present.

> Cost was a major factor in determining what got covered and what did not. To go on a Greenpeace ship, or to join the occupation of the 'Spar and be on the spot, meant that a news editor had to assign an expensive crew with no good idea of when they might be back, or whether anything 'newsworthy' would happen. (Rose 1998, 157)

Thus, 'essentially... they were not there on the spot and so they used Greenpeace video footage' (1998, 156). Of course, the footage presented the issues as Greenpeace saw them and thus the millions of virtual witnesses – on the BBC, ITN (Independent Television News) and other sources – saw on television news bulletins selections from the images that Greenpeace had assembled.

At this point Shell realised that the campaign was going very badly for them. It was hard to rebut the argument about the oil on the buoy since it was one actor's claims against another's – indeed it was one relatively impartial and on-the-spot observer against an apparently more partial and less well-positioned one. And meanwhile any economic advantage to sea disposal was rapidly being swamped by the direct costs of the boycott and by the likely future loss of good name if the image of the company as a selfish polluter stuck. Protestors in Germany threatened direct attacks on petrol stations; two fire-bombings were reported and shots were fired at one garage. Days later, on June 20, Shell abandoned their disposal plans (Thomas and Cook 1998, 7).

Though Shell appeared to be the losers in this about-face, two other actors were also put in jeopardy. First the UK government was left exposed because they had fully supported Shell's disposal method and insisted on its scientific legitimacy. Shell could turn the argument around a little by presenting themselves as a company that listened to its customers. They could say essentially that if their customers did not wish to see the Brent Spar disposed of at sea, the company would respond favourably even if the company's scientists themselves believed that sea disposal was technically the best environmental option. The British government could not so easily present itself as customer-driven. The government had emphasised the importance of non-political, technical decision-making and the following of technical regulations, such as those covering the BPEO, so it found it much less easy to argue that the views of German consumers should be put ahead of the technical decision of the DTI. Chris Rose, campaigns director at Greenpeace in London at the time, also suggests that the UK government was frustrated because they had seen the Brent Spar as central to their strategy for dealing with the wastes from the North-Sea oil industry. Rose argues that the Brent Spar was a handy device for British officials since it did set a precedent for dumping large, metal, even radioactively contaminated material at sea off the Scottish coast: '[this is why] the then UK Government, or at least the DTI, was so furious about the Greenpeace campaign. By stopping the Brent Spar, Greenpeace would seen to have derailed a plan for a submarine rubbish dump on a massive scale, containing not just one "buoy" but perhaps many oil platforms' (Rose 1998, 33). The second actor to come out of this turn of events facing new problems was the environment itself, since it was now no longer clear what would happen to the Brent Spar and all the polluting material it allegedly contained.

The politics of observation

Shell hastily negotiated somewhere to shelter the Brent Spar while the disposal options were reconsidered. In July the Norwegian government granted permission for the buoy to be towed to the Erfjord near Stavanger in Norway, a place which – though rich in wildlife and thus a vulnerable environment – was sheltered and deep enough to keep the installation relatively safe. The efforts by Shell to salvage their reputation were significantly assisted later in the summer of 1995 when Greenpeace admitted that they had made an error in measuring the oil content. Shell had contracted with the Oslo-based marine certification company Det Norske Veritas (DNV) to produce an independent inventory of the buoy's constituent materials, at least in part to reassure the UK government and the public that the company had not been deliberately deceptive about the polluting potential of the platform (see Krebs and Shepherd 1987). According to Rose, in August DNV contacted Greenpeace because of discrepancies between its findings and Greenpeace's claim about the residual oil. DNV's representatives discussed with Greenpeace how the samples had been taken and it turned out that Greenpeace's measurement of the level for its 'sample' was taken not from the storage tanks but from a ventilation pipe, resulting in a false reading of the depth of oil. On 4 September the Executive Director of Greenpeace, UK, Peter Melchett wrote a letter of apology to the Chairman and Chief Executive of Shell, which Greenpeace also made public to the press. Melchett stated that 'We thought samples had been successfully taken from storage tank 1, but we have realised in the last few days that when the samples were taken the sampling device was still in the pipe leading into the storage tanks, rather than in the tank itself'.[3]

In his letter of apology, Melchett is very careful about what he concedes. He admits the mistake but seeks to lessen its impact by claiming that Greenpeace always stated that their results on the oil were 'not definitive'. He also tries to minimise the error by placing its occurrence in context. Thus, in the letter of apology he welcomes the fact that there is now an independent assessment of the contents of the buoy, implicitly criticising the fact that Shell had not commissioned such an impartial assessment early on. Indeed, one could even draw the inference that had there been an independent inventory at the outset, Greenpeace would not have felt compelled to carry out their own assessment and would not therefore have made their mistake. Melchett additionally reasserts the point that Greenpeace's principal objection all along had been to the practice of sea dumping and the danger of setting a precedent. Lastly, in line with Greenpeace's emphasis on the ordinariness of their error, the letter also alludes to scientific mistakes made by Shell, about the marine ecology of the ocean dump site (see below), and invites Shell to make a public admission of their errors too.

The error on Shell's part to which Greenpeace was drawing attention concerned the characteristics of the intended dumping zone. Shell's BPEO case and subsequent arguments had contended that the proposed site was deep and remote. This was important in at least two ways. First, because the site was so far from the ocean surface it was hard to see how pollution from the deep, even supposing there were much, could find its way back near to sea level where life is typically thought to be more abundant and of greater immediate concern to humans. Material in a deep dump would not be pulled up by fishing vessels. And if the fish at depth stay down deep, then any pollution would circulate among bottom-loving biota and not trouble surface dwellers. Second, the suggestion was that conditions in the trough were generally suitable for dumping. The area did not teem with life and it was a quiet, relatively sheltered environment in which the wreck of the buoy would gradually disintegrate and where any contamination would be gradually dispersed.

By contrast, Greenpeace was able to publicise a letter from two marine scientists (John Gage and John Gordon) from the publicly funded Scottish Association for Marine Science (SAMS, see Thomas and Cook 1998, 10; Dickson and McCulloch 1996). The SAMS scientists suggested that Shell's claims were unfounded for several reasons. They maintained that recent ocean surveys indicated that biodiversity in the sea bed may be very high, especially among small invertebrates living in the ocean-floor mud. Accordingly the idea that the dumping would occur in a barely inhabited part of the natural world was mistaken. Moreover the growth rate of such creatures and therefore the level of biological activity is much higher than had formerly been appreciated, so that any idea that pollution might not cause much biological contamination to living creatures appears suspect. The scientists at SAMS also pointed out that, with developments in fisheries, more deep-water fish – such as orange roughy – are fished, even down to 1800 metres. And many of the fish species caught at depth are also known to visit levels much closer to the surface so that any idea that the ocean-floor is biologically segregated is not tenable. The intended disposal site was also close to a sea mount (a submarine mountain), the sides of which are attractive to fish so this site is likely to be relatively well stocked with fish moving from one stratum of the ocean to another. Lastly, the proposed site is not, as appears to have been thought by some, an area of tranquillity. It is in fact susceptible to turbid 'storms' that can sweep the sediment up at high speeds. Any suggestion that dispersal of pollution from the Brent Spar would necessarily be gradual thus appears highly suspect.

In such ways, Greenpeace tried to find what little virtue they could in this unhappy turn of events. They stressed their own honesty and contrasted this with the chronic reluctance of other institutional actors to own up to their errors. But it was too easy for their critics to discern a pattern: Greenpeace appeared to have made exaggerated claims at the height of the dispute which they safely retracted once the decision had gone in their favour.[4] Greenpeace representatives were adamant that it was not how events unfolded. But their political opponents were not inclined to give them the benefit of the doubt. The UK government turned on them immediately for opportunism. Much of the British press followed suit. And once the DNV inventory analysis appeared in mid-October, Shell was able to emphasise the similarity between much of the impartial assessor's listing and their own, further reinforcing their public stance that they were acquiescing to the public's wishes not going back on their technical analysis.

Shell and the management of the Brent Spar's disposal

Over the ensuing years Shell was able to maintain a dual position. Without ever admitting that its science had been incorrect or that its original dumping plans constituted anything but the most effective disposal solution, Shell engaged with apparent enthusiasm in the drawn-out consultative process that led to the ultimate dismantling and partial reuse of the buoy. Given that the company's public stance was that it was adopting a different disposal method in response to customer- or public-demand, they had no need any longer to themselves identify the best disposal option. Instead, they retained DNV and other consultants to run a lengthy international consultative process in which a number of alternative disposal options were reviewed and concrete plans for the buoy's future examined (Rice and Owen 1999). The meaning of the 'best' disposal option subtly shifted from something that technical experts (ideally) would identify, to the option that was widely deemed to be the most acceptable. In the end, the topmost structures were removed from the Brent Spar and the rest was divided into large cylindrical sections – effectively cut into slices. These slices were transported to a nearby Norwegian port, Mekjarvik, where the sections were installed in the sea a few metres from the shore and used as the base for a new platform that made up a quay expansion and ferry terminal. Of course, this solution too required a BPEO. But given the international openness of the procedure and the transparency of the assessment of the pros and cons of all the suggested disposal options, British officials readily approved the scheme proposed. The plan was agreed in August 1998 and the work completed by early 2000.

Throughout, Shell kept itself at a distance from claims about the correctness of the judgements surrounding the new disposal option.

Though it insisted that its initial inventory was largely accurate and that the original BPEO was robust and objectively defensible, the company now insulated itself from the analytic work involved in the disposal process. In a sense it was no longer concerned with what was correct; it was more concerned with arriving at a solution that was transparently chosen and widely agreed. The goal had moved from objective correctness to acceptability.

Conclusion: truth, trust and observation

The case of the Brent Spar demonstrates powerfully Shapin and Schaffer's insight that observation is indexically tied to trust and to institutions for the generation and maintenance of trust. The gentlemanly problems faced by Robert Boyle reasserted themselves in the dispute over this storage buoy. Without a reason to trust the observer, observational truths are powerless. First, Shell's claims about the buoy's constituents and then Greenpeace's observational counter-claims about the oil came to be doubted. On the high seas, the truths of the Brent Spar were inscrutable since there were no other witnesses to testify to the correctness of the observations. Observational 'truths' turned out to be highly suspect, in the first case because Greenpeace produced grounds for distrusting Shell's veracity as an observer and, in the second, because Greenpeace's observational claims were based on unskilled and defective observational practices.

Moreover, there were very few witnesses actually present. Only parties to the original BPEO were involved in the first inventory estimates and only Greenpeace personnel were involved in the direct oil measurement on the installation itself. Ironically, the millions of virtual witnesses offered by television news coverage of the state of affairs on the Brent Spar turned out to be a problematic resource in the warranting and legitimation of scientific facts. It may be that, as the quote from Shapin indicated earlier, 'Virtual witnessing involved producing in a reader's [or viewer's] mind such an image of an experimental scene as obviated the necessity for either its direct witness or its replication'. But it turned out that the virtual witnesses were not authoritative allies since they too were inexperienced and inexpert regarding the difficulties of measuring the amount of oil on the buoy.

This situation was made even worse when the suppliers of the television news, for their own reasons, turned their critical fire on the quality of the observations. As Rose notes at some length, once the media companies found that they had swallowed the misleading Greenpeace line on the amount of residual oil, they tended to blame Greenpeace without publicly assuming any responsibility themselves. Rose quotes David Llyod, then senior commissioning editor of Channel 4's News and Current Affairs,[5] speaking at the 1995 Edinburgh Television Festival and reflecting on the furore surrounding Greenpeace's admission of their error:

> On Brent Spar we were bounced... By the time the broadcasters had tried to intervene on the scientific analysis, the story had been spun far, far into Greenpeace's direction... When we attempted to pull the story back, the pictures provided to us showed plucky helicopters riding a fusillade of water cannons. Try and write the analytical science into that to the advantage of the words. (Rose 1998, 158)

From Rose's point of view the news media were not 'bounced'; they had only themselves to blame since it was their lack of independence that led them to rely on Greenpeace's assessments. Nothing but finance and convenience stopped them from making their own estimation of the oil content. But they blamed Greenpeace and, in lectures and presentations at the television festival in Edinburgh, the TV industry made exculpatory claims about their serfdom.

> After the Edinburgh TV Festival the BBC told the *Financial Times* that: 'We suppose to a certain extent we illustrated at Edinburgh that there was an attempt to manipulate a situation by Greenpeace. They've now admitted to a situation which demonstrates how important it is for the media to test the veracity of what any organisation declares or states'... So the error over the oil, which Greenpeace freely admitted to, was used to try and cement an argument that the televisual media had been misled. The implication was that – somehow – what the pictures showed during the course of the occupation and the tow of the 'Spar essentially misrepresented the reality. But how? It was certainly not to do with the mistake over the oil – which was a mistake and not a manipulation of data or an invention. (Rose 1998, 160–61; see also Hansen 2000)

Rose is at pains to point out that the error in estimating the oil was independent of the visual coverage and that the mistake over the oil came late in the dispute, when the tide of political and popular opinion had already turned against Shell. His principal concern seems to be that Greenpeace's admission of their error became a justification for the leading UK news media to adopt a persistently sceptical attitude towards environmentalists' (and specifically Greenpeace's) claims. But Rose's resentment also shows to what extent observation is a moral and political business. Observations are not made in an individualistic vacuum; they depend for their convincingness on a context of mutual trust. Greenpeace realised that their observational error had become a licence for commentators to withdraw trust on future occasions. Greenpeace could no longer count on being lone speakers of the truth since:

> This sort of individualistic rhetoric, taken by itself and at face value, would count as a massive misrepresentation of scientific practice. In fact, seventeenth-century English natural historians and natural philosophers, writing in other moods and for other purposes, showed themselves well aware that it was. Many of the same practitioners who produced some of the most vigorous individualistic methodological pronouncements also displayed keen appreciation that there was a proper, valuable, and ineradicable role for testimony and trust within legitimate empirical practices. (Shapin 1994, 202)

After Brent Spar, the challenge for Greenpeace was to re-authorise the forms of testimony on which their campaigning activities depend.

10
Genetically Modified Organisms and the Unbearable Irresolution of Testing

Introduction: sustainable development and environmentalism-as-usual[1]

The early 1990s appeared to offer a key turning point in the development of environmentalism. Endorsed by the Earth Summit in Rio in 1992, the concept of sustainable development seemed set to transform activism around environmental issues from an outsider's critique of advanced-industrial society to a central component of the officially sanctioned solution. Environmentalists could come in from the cold because all development now had to be sustainable; this meant that everyone – governments, industry, international bodies and NGOs – should be pulling together for the same objectives. Governing parties the world over were authoring documents charting their paths towards sustainable development. In principle at least, environmentalists were indispensable to plans for a sustainable society and were to be offered a key role in the reform and re-development of core socio-economic activities.

At the same time it appeared that environmental problems, even the most intractable ones, could be tackled. The hole in the ozone layer had been acknowledged and was now likely to heal itself in the coming decades. Institutional arrangements for recycling and waste reduction advanced beyond previous expectations. Advocates of ecological modernisation argued that firms could benefit economically from cutting emissions and waste (see Mol 1997). The nuclear industry was in retreat and the scope for renewable energy looked attractively large. It began to seem that environmental management was all about balancing trade-offs, and the language of environmental economics became the dominant mode for describing such trade-offs with optimum rationality (even among many NGOs). In this context, climate change occupied a key role. Clearly the problem was vast, uncertain and complex. Huge vested interests stood in the way of arriving at solutions before excessive and perhaps irremediable warming took place. But in the run-up to, and the immediate aftermath of, the signing of the Kyoto Protocol in December 1997 there was still a cautious optimism that a solution could be found. The Intergovernmental Panel on Climate Change (IPCC) had been established to provide an international forum for establishing reasonably certified scientific claims about the scale and nature of the problem, and negotiations began about how the burden of coping with the issue could be shared in a practicable but also reasonably equitable way. While it would be wrong to say that environmentalists and policy analysts were comprehensively optimistic, their pessimism was at least bounded.

Some unease disturbed this rough consensus. The Brent Spar marked a ferocious though short-lived controversy even if, in the end, Shell managed to find a way of acceding to the environmentalists' demands. Also in the UK, in the mid-1990s anti-roads protesters argued that governments were not treating environmental protection anything like seriously enough within transport planning. Indeed, many believed that 'sustainable road-building' was oxymoronic. But even these arguments – played out with innovative techniques such as tunnelling beneath the path of the diggers and the creation of new dissenting heroes as noted in Chapter 2 – could still be handled within the dominant discourse of sustainable development. The problem was that governments were failing to implement their in-principle commitment to sustainable development in the case of road transport; they simply needed to be embarrassed back into line. In North America, similar issues came to the fore with respect to wildlife conservation and forest management and exploitation.

By the end of the millennium, however, this temporary alignment of forces in the UK around the concept of sustainable development had been ruptured; environmental campaigning had returned to its oppositional status and Peter Melchett of Greenpeace was (however briefly) imprisoned. Genetically modified (GM) foods were the occasion for this change.

The promise and threats from genetically modified organisms (GMOs)

As is well known, genetic modification refers to the ability to introduce novel traits into a species of plant, animal, bacterium or virus by inserting alternative genetic material. The technique has many possible fields of application: bacteria might be developed which could consume hazardous wastes producing only inoffensive by-products; animals might be bred to yield medicinal products needed by human patients; plants could be developed, which might generate medicines as well as their usual fruits; and crops could be produced which made agriculture more straightforward. To date, there has been less antipathy to the medical procedures. Rather, concern has focused on the agricultural products.

There are currently different kinds of agricultural biotechnology innovations on offer. Farmers have been changing the character of crop plants for millennia, by selectively retaining advantageous mutations and by deliberately crossing plants with desirable characteristics. More radical steps have also been taken to promote mutations, through for example subjecting plants to severe chemical treatments or exposing them to radiation. These interventions were unpredictable and thus the plant breeders still had to wait to see what the character of the mutations might be. The biotechnology industry presents its activities as only an enhancement of such time-honoured procedures with the added benefit that the changes are not random and can be – approximately – anticipated in advance.

The industry and many governments have presented several kinds of advantages that may be expected from these procedures. Some are rather speculative, such as plants that are drought resistant or which produce medicinal products. Others are more tangible: the market has been offered fruits with longer storage lives, plants with in-built pest resistance, and plants that are resistant to proprietary weedkillers. In each of these latter cases there are said to be various resulting benefits: some are market advantages, others broadly socio-economic or environmental. Thus in the case of the first of the above innovations (longer-lasting fruits), the benefit claimed is that there is less wastage, a jointly economic and environmental bonus. In the second case, exemplified by 'Bt-maize' (a maize, or corn, engineered to produce the protein created by the *Bacillus thuringiensis* bacterium[2]), the argument is that the crop is given a natural resistance to corn-boring pests. This should both benefit harvests and reduce the use of expensive and environmentally disadvantageous pesticides. The Bt-gene derives from a bacterium that produces a protein which the corn-boring larvae find toxic. Existing (that is, non-GM) treatments derived from this Bt source are recognised as organic. Enthusiasts for biotechnology accordingly argue that this approach is more benign than common agrochemical treatments for the problem in maize. It uses a 'natural' genetic process to yield pest resistance, eliminating the need for chemical pesticidal sprays. The third case is, to date, the commonest pattern. A company produces crops with a gene that conveys resistance to the company's own weedkillers. In principle, this means that the farmer can spray the whole field with the weedkiller at key stages in the growth cycle, eliminating all the weeds but leaving the crop unaffected. Previously, such blanket applications would have been impossible since the herbicide would have killed the crop as well, except in those rare cases where a 'natural' resistance to the weedkiller had arisen in the crop. This process is supposed to boost production (as weeds cannot impair the crop) and reduce the amount of spraying that needs to be carried out, since the spray is now more effective. So efficacious should the weedkillers be that the spraying can be undertaken at a time which suits local conditions. This should be beneficial for wildlife conservation as it could – for example – allow nesting birds a longer uninterrupted time in the fields before the spraying had to be undertaken.

Objectors dispute nearly all these claimed advantages. Part of the objection is at a general level. It is suggested that plant breeding to date has used variations which occur naturally while biotechnology introduces genes from widely differing biological realms. Of course, plant material has been treated in pretty 'unnatural' ways by plant breeders to date. Still, there are limits on the crossing that can be attempted with conventional plant breeding. By contrast, the new technology allows genes from bacteria to be inserted into plants, or those from fish into fruits. While the imported genes typically still produce the traits for which they 'code' in the parent species, it is unclear what other consequences may arise from the genetic manipulation. This causes anxieties both in relation to the so-called environmental fate (that is, what might happen to the species in the wild) and in terms of food safety. While this latter objection is, in a sense, necessarily speculative since people have not been eating GM food for long, there is some experimental evidence from the notorious and hotly disputed tests conducted in Scotland by Dr Arpad Pusztai then of the Rowett Research Institute. He fed rats with potatoes which had been genetically modified to produce a snowdrop lectin, a protein that might promote insect resistance in foodcrops; he also fed a control group regular potatoes 'spiked' with lectin derived from ordinary snowdrops.[3] Those eating the GM form appeared to become sicker, possibly because of a problem with their immune system, though none of the rats did well on this restricted diet. As described below, the handling of these experimental results became enormously contentious. They do however illustrate the diffuse worry entertained by many – that the biological implications of genetic manipulation may be more complex than the proponents of the technology commonly acknowledge. At a more mundane level, tests on soya which had received genetic material from Brazil nuts reported the soya to contain the very substances by which consumers who are intolerant of nuts are affected. The resulting fear is that the soya retail products would not have been labelled as 'containing nuts' so that health problems could easily have arisen for consumers of the GM soya. For opponents of the technology, this example is attractively concrete. However, it does not appear to give proof of novel, unexpected health threats arising from genetic modification since the sufferers would have been expected to react to Brazil nuts in any event. In a sense, this danger was rather predictable and thus not a good exemplar of the kind of unanticipated hazards that many critics like to focus on.

A second set of concerns relates to the environmental consequences of GM material. Though this issue appears to be about a single thing – environmental fate – it is actually more complex than this. For example, one potential concern is that GM foodcrops could cross with biologically related weeds, allowing the introduced gene to enter the weed population also. A weedkiller-resistant crop plant might then give rise to weedkiller-resistant weeds, and so on. Crops could also cross with other, related crops thus spreading genetic characteristics in an unplanned and untested way, with possibly undesirable long-term consequences. Secondly, it is possible that the pests targeted by GM-related farming methods could develop resistance to those techniques. For example, corn-boring larvae that exhibit any Bt-resistance will have an enormous evolutionary advantage over those which do not, threatening to change the nature of the pest itself. Similarly, vast monocultures of field after field of weedkiller-resistant crops would be an ideal setting for any weeds which naturally developed a similar resistance. But environmental fate is not restricted to these specific mechanisms. Campaigners made much of a US report on experimental data indicating possible damage to monarch butterfly larvae resulting not from them feeding on GM foodcrops (the larvae are not interested in maize), but from them ingesting pollen from GM crops which could drift onto neighbouring wild plants that they did eat (*ENDS Report* 292, May 1999, 28). In this case, the damage was wholly 'collateral'.[4] Critics of GM farming also argued that, even if herbicide-resistant crops should allow growers to use fewer sprays, the temptation would be to overuse herbicides knowing that the crop was impervious to treatment. Rather than encouraging a regimen with fewer agrochemicals, GM plants might stimulate extra spraying, with possibly deleterious effects on the environment and on the health of consumers because of agrochemical residues on the food. Furthermore, critics argued that wholly weed-free fields would not necessarily be a desirable thing as such fields would be wildlife 'deserts'; for example, elimination of weeds could lead to the disappearance of the insects that are preyed upon by birds, making the countryside increasingly inhospitable to avian species.

One further consumer factor was very important, though in a sense rather incidental to the overall GM enterprise. In developmental work on gene transfer, plant technologists had wanted an easy way to check whether genetic material had been correctly incorporated. Since the characteristics they had introduced would only be displayed once the organism had matured, they sought a means of establishing sooner whether the gene had 'taken'. Accordingly, a segment of genetic material which gave rise to antibiotic resistance was inserted alongside the other transferred genes as a marker. The new material could be checked for antibiotic resistance early on and if that resistance existed it was safe to assume that both the marker and the desired gene were in place. In the haste to get these GM products on to the market, the antibiotic-resistant gene was not taken out, so that fodder and foodcrops with built-in resistance to antibiotics came on to the market. At a time when there was widespread concern in the public health sectors about the excessive prescribing of antibiotics and about routine use of antibiosis in intensive livestock-rearing (both of which appeared likely to stimulate the evolution of antibiotic-resistant infections) a new source of antibiotic resistance in the food chain seemed particularly undesirable.

Two other broader concerns were also raised repeatedly. The first had to do with organic production. Across the advanced industrial world, interest in organic produce is increasing. By and large the organic movement has been opposed to genetic modification. Rather inventively, the US plant technology industry initially attempted to argue that since gene transfer simply moved 'organic' material from one plant (or bacterium and so on) to another, GM foods could themselves be regarded as organic. An attempt was made to get the federal regulatory authorities to include GM foods (grown organically of course) within the official organic label. This effort was defeated. But organic growers still face the problem that if, say, they are cultivating a maize crop of some sort close to fields of GM maize then their crops may cross, allowing the genetic modification into future generations. The potential for this to happen was graphically illustrated at an English test site in Devon in 1998 where experimental cultivation of GM maize crops took place very close to neighbouring (and as it happened organic) fields in which sweetcorn was being grown; the distance between the GM fodder maize and the fields where the organic farmer could have planted was only 275 metres, though in the end organic planting was kept about 2 kilometres away. ACRE (the official advisory committee on releasing trial species to the environment, which advises the Environment Ministry, see p. 169) advised that even 200 metres would have been a reasonable separation, although they expressly disregarded organic farmers' own notions of acceptable levels of purity in arriving at this estimation.[5] The second concern has some affinities with the ethos of the organic movement, focusing centrally on the ownership of seeds and of the intellectual property they embody. To make profits for industry, GM crops have to be patentable. In this way, farmers can only buy Roundup-Ready seeds from Monsanto, the manufacturers of the weedkiller Roundup, and so on. Critics argue that this has negative consequences, both for the developing world where farmers save their seeds for next year's planting and for the organic and other seed-saver movements in the advanced industrial world. On this view, GM cultivation becomes the Trojan horse allowing monopoly control over seeds. Even if GM technology were to produce foodcrops with desirable characteristics for developing-country agriculture (with in-built drought resistance for example), farmers will be drawn further into the cash nexus. They would not be the lawful owners of the right to the seeds from this year's crop but would instead have to buy new seeds each year. A further irony is that the source of many of the genetic materials is the biodiversity-rich areas of the world where – at least in part – biodiversity has been maintained by non-industrialised farming and by the practices of seed-saving and smallholder cultivation. Disputes have continuously arisen over the division of intellectual property rights between the companies which 'prospect' for the beneficial genes and the farmers and other locals whose cultivation methods and way of life happen to have resulted in this intellectual property being available.

The nested hierarchy of authority and credibility in the GM debate

This case has important similarities as well as important differences from the Brent Spar dispute considered in the last chapter. In common is the fact that both controversies demanded that environmentalists and the authorities make judgements about scientific and technical issues; both required the weighing of precautionary concerns. Both controversies saw objectors and proponents argue the case out through the language of science. Scientific expertise was needed to work out whether dumping the Brent Spar in the Atlantic would be highly damaging to the environment or not. In this case, scientific understanding is required to figure out what implications field upon field of GM fodder maize might hold for wildlife. In the Brent Spar case one had to make precautionary judgements. If one sank the buoy, would it break up on descent or would it glide smoothly to the sea floor? In the case of GM crops too, one has to decide how long and how broadly to test the food safety of GM crops with built-in insecticidal properties. But the difference is also important. The Brent Spar was one oil platform. Though its disposal might have set a precedent for aspects of the oil industry, sea dumping is already highly regulated and subject to international agreements. In the GM case one is considering whether or not to introduce a novel farming technology that will most likely have irreversible implications across the globe, as described in Chapter 4. The demands for a scientific assessment of the issues involved and the weight attached to judgements about exactly how much precaution to exercise are more acute in the GM case. It turns out that the demands for impartial and expert scientific judgement and for an agreed interpretation of precautionarity simply could not be satisfied and that appeals to higher authorities brought no resolution either.

From the outset, it has been clear that some conceivable dangers attach to this new technology. Proponents have tended to acknowledge that there are conceivable risks associated with the new crops but argue that no technologies are risk-free and that current farming practices are risky in certain senses also. Of course, they indicate, GM crops cannot be guaranteed free of risk but the putative risks (to consumer health, to beneficial insect life and so on) have been investigated as exhaustively as is practically possible and there is no indication that the technology is specifically risky or likely to be more harmful than the way farming is carried on at the moment. No particularly sophisticated analysis of the nature of scientific reasoning is needed to understand the problems to which this leads. In what one may as well call a straightforward Popperian way, it is clear that one simply cannot prove a negative assertion – one cannot *prove* the absence of possible harm. Attempts to break out of this difficulty have often invoked the notion of the precautionary principle as discussed in Chapter 8. But rather than resolve the problem, the switch to the language of precautionarity has tended to entrench the divide. Just as opinion is divided about GM crops, so views are polarised about how much precautionary testing one needs to undertake before one can reasonably authorise their use. Given that adverse environmental consequences could be several generations away or that the development of weedkiller-resistance might take years to arise, it is not clear how many seasons' testing would count as good evidence of safety. The precautionary principle alone says little about how much to err on the side of caution (Levidow 2001); those with opposing views of GM technology interpret the precautionary principle in conflicting ways. Initial tests by the Department of Agriculture in the USA set the level for evidence of harmlessness rather low. Once the technology was adopted in North America, consumers were effectively taking part in a large-scale trial. After a few years passed, other nations' official agencies have been able to imply that upwards of half a decade of testing on a large North American population is precaution enough. Environmental campaigners suggest that this is but a meagre test, though they are generally careful never to say how long a wait precautionarity is likely to demand.

Attempts to refine the standard of proof have only led to further disagreement. Discordant interpretations of precautionarity have taken a more precise form in disputes over the standard known as 'substantial equivalence'. As Millstone *et al.* pointed out in a contribution to the journal *Nature* (1999), in order to decide how to test the safety of GM food, one needs to make some starting assumptions. Precisely because GM crops are – by definition – different to existing crops at the genetic level, one needs to decide at what level one will begin to test for any differences that might give cause for concern or even rule the new crop-technology out. According to Millstone *et al.*:

> The biotechnology companies wanted government regulators to help persuade consumers that their products were safe, yet they also wanted the regulatory hurdles to be set as low as possible. Governments wanted an approach to the regulation of GM foods that could be agreed internationally, and that would not inhibit the development of their domestic biotechnology companies. The FAO/WHO [UN Food and Agriculture Organisation/World Health Organisation] committee recommended, therefore, that GM foods should be treated by analogy with their non-GM antecedents, and evaluated primarily by comparing their *compositional data* with those from their natural antecedents, so that they could be presumed to be similarly acceptable. Only if there were glaring and important compositional differences might it be appropriate to require further tests, to be decided on a case-by-case basis (1999, 525, emphasis added).

Regulators and industry agreed on a criterion of substantial equivalence as the means for implementing such comparisons.

By this standard, if GM foods are compositionally equivalent to existing foodstuffs they are taken to be substantially equivalent in regard to consumer safety. Thus, GM soya beans have been accepted for consumption after they passed tests focusing on a 'restricted set of compositional variables, such as the amounts of protein, carbohydrate, vitamins and minerals, amino acids, fatty acids, fibre, ash, isoflavones and lecithins. [They] have been deemed to be substantially equivalent because sufficient similarities appear for those selected variables' (1999, 526). However, as Millstone *et al.* argue, with just as much justification, regulators could have chosen to view GM foodstuffs as novel chemical compounds coming into people's diets. Before new food additives and other such innovative ingredients are accepted, they are subject to extensive toxicological testing. These test results are then used very conservatively to set limits for 'acceptable daily intake' (ADI) levels. Of course, with GM staples (grains and so on), the small amounts that would be able to cross the ADI threshold would be commercially insufficient. But safety concerns would be strongly met. These authors' point is not so much that GM foods should be treated as food additives or pharmaceuticals, but that the decision to introduce the substantial equivalence criterion is not itself based on scientific research. That decision is the basis on which subsequent research is done. For proponents of the technology, substantial equivalence is a straightforward and commonsensical standard. If GM soya has the same constituents as regular soya then it can surely be eaten. But the standard conceals possible debate about what the relevant criteria for sameness are. As Millstone *et al.* point out, for other purposes the GM seed companies are keen to stress the distinctiveness of their products. GM material can only be patented because it is demonstrably novel. How then can one be sure that it is novel enough to merit patent protection but not so novel that differences beneath the level of substantial equivalence may not turn out to matter a decade or two into the future?

As we have seen, neither the appeal to the precautionary principle nor the adoption of the substantial equivalence standard provides the basis for agreement about the way in which the safety and environmental suitability of GM crops can be authoritatively assessed. But this interpretative unravelling of the very devices introduced in order to resolve the problem continued; from abstract principles it moved on to the composition and character of regulatory institutions. European governments, like their US counterparts, have often appeared to be supporters of the agricultural biotechnology industry. Among the first GM crops to pass European regulatory hurdles was a French-Swiss maize. France, Spain, Italy, the Netherlands, Portugal and Denmark have large agricultural sectors and their governments are naturally interested in the well-being of this sector. Centre-left governments in France and Britain, for example, in the second half of the 1990s were keen to demonstrate their business-friendly credentials; this was the heart of the so-called Third Way. Industry and agricultural ministries did not want to see the new biotechnology industry surrender to the technical advances of US corporations as had appeared to happen in the IT and computing area. But then, as it became apparent that the methodological 'principles' intended to guide the assessment of GMOs depended to a very large extent on interpretative judgements, attention switched from the principles themselves to the make-up of the bodies charged with doing the judging. In Britain, ACRE had many members with links to the industry and only one representing environmental and consumer interests. It was easy to suppose that commercial considerations could influence the way the judgement was made.

Such partisanship was made to seem all the likelier by other recent episodes. Most obviously there was mad-cow disease, bovine spongiform encephalopathy (BSE). It was clear that, from its earliest identification in the mid–late 1980s, there had been great uncertainty about the nature of and the risks from BSE. The British (and other European) government faced a dilemma. Kindly put, their problem was this: did they accept the worst risk scenarios, declare a possible danger to human health from beef-eating and thus precipitate the collapse of the beef and, maybe, dairy industries, or did they use the fact that there was no clear evidence of cross-species transfer to say that there was very likely no risk to consumers and thus create the opportunity to reform the industry gradually. The UK government opted for the latter path and were led to make more and more public pronouncements to the effect that the risks to human beef-consumers were very slight or minimal. In 1993 the then Environment Minister publicly – and very memorably – fed a beef-hamburger to his young daughter, indicating the confidence he felt in the safety of British beef and in the tendency of spongiform encephalopathy to respect species boundaries. Three years later the same Conservative government came to admit that there was persuasive evidence that humans did risk infection from cattle meat; the worth of the assurances about safety slumped to zero.

Though the appellation 'mad' came to be applied to some GM crops, particularly in France where all manner of foodstuffs were captioned 'fou', the analogies between the two cases are limited. Aside from Pusztai's contentious potatoes, there was none of the evidence of health risks which were so apparent from the disturbing behaviour of mad cattle. 'Mad' maize is not conspicuously madder than its sane relatives. But it was apparent that the conflict of interest was fundamentally the same as in the BSE case; and the possibility that commercial considerations would affect the weighing of the risk assessment was also very much the same. In both cases there were expert panels advising the ministers, but both panels were composed of invited experts. Ministers determined whose advice they listened to. The same secrecy and confidentiality surrounded the cases. Even if the advisory panels did include the best experts, there was no guarantee that they had a precautionary version of the public interest at heart.

Finally, in the GM case it began to appear to some opponents of the GM technology that the scientific establishment in Britain was *parti pris*. Under ideal (though possibly unrealistic) circumstances, the scientific community should be offering impartial and disinterested advice on this matter, as on global warming or ozone depletion. However, in the response of the Royal Society in London to Pusztai's experiments, some saw a lack of impartiality. As mentioned above, Pusztai was a career scientist at the Rowett Research Institute in Scotland. This Institute, near Aberdeen, carries out research sponsored by many agencies and firms, including some companies involved in the GM business. When his early experimental work on the GM potatoes became public through the mass media, Pusztai was criticised by his own Institute on the grounds that he should not have publicised studies that were preliminary, unchecked and, therefore, possibly mistaken. Only once the results had been authorised by the process of checking carried out by other scientists, known as 'peer review', and published in a scientific journal should he have been willing to discuss them with the popular media. Up to that point, his findings were private and unconfirmed; their factual status had not been attested. Pusztai lost his position for disregarding rules about responsible publication.

This verdict was tough, particularly since other scientists published work in broad agreement with his studies and his own study was finally published as a paper in the reputable journal *The Lancet*. Still he had gone public with unconfirmed findings in a contentious area so some disciplinary action was unsurprising. What was more surprising was that his experiments were subject to a lengthy and critical review published by the Royal Society in London.[6] This was unusual because flawed work is normally just ignored by the scientific community and left to sink without trace. The Royal Society does not spend its time rooting out bad science and heaping opprobrium on the publishers of poor work. That the Society had chosen to do so in this case suggested an asymmetry in their approach. They objected to 'scientific' studies that could be used to oppose technological progress but did not bother to condemn scientists' over-sanguine claims about the benefits of novel food technologies. The scientific establishment appeared to see popular opposition

Table 10.1 Levels of contention in the UK GM debate

Procedural issues	• Showing the absence of harm • Deciding how precautionary to be • Interpreting substantial equivalence
Institutional issues	• Deciding who should select the experts • Impartiality or partisanship of the scientific establishment

to GM crops as scientifically unsupported. They tried to speak up for science but in doing so seemed to speak up for GM.

The argument of this section is summarised in Table 10.1. On the face of it, pro- and anti-GM groups agree that scientific judgement is needed to assess the safety and advisability of accepting GM foodcrops.[7] But at each stage at which that assessment is attempted, the advocates and opponents disagree. They find no agreement over how to show the absence of harm. If that issue is recast in terms of the precautionary principle, then the degree of precautionarity is itself disputed. The standard of substantial equivalence was introduced to overcome problems with the interpretation of the precautionary principle but that standard is disputed also. Moving from procedural matters to institutional ones brings no relief either since the government's choice of scientific experts is itself presented as suspect while the ideal of a disinterested scientific community appears to be thrown into doubt by the handling of the Pusztai incident.

Playing out the dispute in the absence of proof

In the absence of scientific proof about the safety or riskiness of GMOs – indeed in the absence of any agreement about what 'proof' might amount to – farmers, consumers, retailers and agricultural supply companies still had to get by (see Levidow 1999). Individual consumers and other food providers, such as schools, hospitals and local authorities, turned elsewhere than to the regulatory and scientific authorities for assurance. Given the dominance of supermarkets in the food-retail sector in Britain, supermarkets took a lead role in devising methods for responding to public anxieties. The trend was started by a small chain that specialised in frozen foods, Iceland. In 1998 they undertook to eliminate all GM ingredients from their own-brand products. Other supermarkets rapidly followed suit and the leading stores, which were already involved in a titanic struggle for market share, used this new weapon to try to gain a little edge over their rivals. The government was even called on to aid these large retail companies, with civil servants from the trade ministry providing advice on where certifiably non-GM ingredients could be obtained. Of course, it was practically impossible for consumers to tell whether materials had been sourced from GM-free supplies, but the supermarkets met with little of the scepticism routinely directed at scientists and official regulators. Apart from the occasional accusation that, for example, underwear from Marks and Spencer (a supermarket-cum-department-store) was not made from GM-free cotton and thus the store was not fully committed to protecting customers from buying GM goods, supermarkets appeared to handle the issues of cognitive authority successfully. Essentially they used techniques of reassurance which had been built up during the height of the bovine spongiform encephalopathy (BSE) crisis, stressing the personal contacts between supermarkets and their suppliers, the dependable and enduring nature of these connections, and even featuring profiles of particular farmers on packaging (on strategies in the BSE case see Jasanoff 1997). And, in concert with other EU states, the UK government acceded to demands for labelling of food products made with GM ingredients, even in restaurants and other public facilities (see Chapter 4).

Though several crops had now passed all relevant European tests and were therefore legally cultivable in Britain, the industry and the government recognised that 1998 and 1999 would be inauspicious times to introduce them to British agriculture. Resisting calls for a moratorium on GM crops – calls supported by NGOs and some influential newspapers – the government decided on a policy of a limited number of farm-scale trials. The idea was to grow a variety of crops – maize (in the US, corn), fodder and sugar beet, and oil-seed rape (in the US, canola) – on volunteered farmland over four years (1999–2002) at the field-level scale and then to make an assessment of the environmental implications on wildlife, neighbouring fields and so on. Ostensibly, the hope was that the environmental effects would be shown to be negligible and therefore public anxiety would be allayed. However, NGOs – which had already pursued a policy of occasional sabotage of GM trials – took direct action against many of these farms. In line with recent (unenthusiastic) steps towards open governance, the UK government published full map references for the test sites on its websites. No police force was willing to pay for round-the-clock protection of remote agricultural locations, thus handing a sure victory to any would-be saboteurs. And so a series of raids on the test plantings began. A variety of justifications for such action were offered: some protesters adopted the familiar

EarthFirst! line to the effect that it is the role of protest to simply halt undesirable actions. One drives up the cost of the (allegedly) environmentally damaging action so that the perpetrators think twice about doing it. In this case, if the government had committed itself to a position of not approving cultivation until farm-scale trials were completed, indefinite delay of the trials must mean that cultivation would never begin.

Even the actual purpose of the field trials was disputed. Greenpeace spokespersons were inclined to argue that the trials, which were relatively few in number and dealt with different crops so that the sample size of any particular crop-type was very small, had little scientific rationale. On this view, the plantings were more reasonably interpreted as an exercise in 'softening up' public attitudes by producing a GM fait accompli. Of course, Greenpeace would not have welcomed the additional cultivation sites which would have been needed to increase the sample size. But, as though in support of the Greenpeace interpretation, some of the trial plantings which were supposedly justified in the name of science appeared to be amateurish. The GM crops were planted too close to related foodcrops or were inadequately inspected; Friends of the Earth (FoE) mounted legal challenges on these kinds of grounds.

At the same time, Greenpeace and other campaigners had no interest in precision in their public opposition to GM plantings. They appeared happy to conflate anxieties about health and about adverse environmental effects. It is clear that they needed to avoid making any concessions to the industry for fear of 'Trojan Horses' so all GM plantings had to be interpreted as hazardous. By way of illustration, in July 1999, 28 Greenpeace protesters arrived in the early morning at a trial site in the east of England intending to 'decontaminate' a field of GM maize. For publicity purposes the protesters wore protective overalls-style suits, as though simple contact with the GM corn might be dangerous. The three farming brothers who owned the farm were more alert than the environmentalists had supposed and ambushed the action, ramming the Greenpeace equipment with their tractors and pinning the Greenpeace transport down. The Greenpeace activists were arrested. Just over a year later the case came to final trial and the activists were acquitted by the jury. Jury decisions in the UK are confidential so no commentator knows the basis of the jury's decision; however, given the apparently incontestable evidence of criminal damage, it seems likely that the jury decided that they were in sympathy with the action. Greenpeace got little opportunity in court to test arguments about the 'contamination' and 'pollution' from the crop since the prosecution chose to focus only on the criminal damage perpetrated. Though the case did not set a precise precedent, officials and farmers feared that future acts of sabotage would likely also evade prosecution; one political cartoonist drolly suggested new signs be erected in fields announcing that 'trespassers will be acquitted'.[8]

In addition to the field trials, the Labour government introduced one further initiative intended to provide an additional source of evidence, a public consultation exercise known as 'GM Nation'. This type of exercise had been proposed by an advisory group to the government, the Agriculture and Environment Biotechnology Commission; in 2003 the government provided funds for it to run. Public consultations were held in six centres across the UK and views could be submitted online also. Views were synthesised into a report that emphasised the depth of disquiet about the advisability of GM crops and a general antipathy towards early commercialisation. However, given the design of the consultative exercise, people with strong views tended to be over-represented among those who gave opinions. Moreover, the exercise was designed to be entirely consultative. The government ceded none of its decision-making authority to the consultation, although of course large-scale opposition voiced through GM Nation might well act as a political obstacle to the government speeding, for example, towards commercialisation. Neither GM Nation nor the field trials therefore are likely to provide the independent yet overwhelming evidence that would allow an agreed course of action to be identified.

Conclusions

The GM case points to an irony. Scientific expertise is needed, everyone accepts, to make judgements about novel technologies and their consumer safety and environmental advisability. But beyond that point the agreement fades. What the scientific tests should be, what forms of evidence should be sought, who should interpret the scientific evidence? – these are all issues that remained unresolved in the GM case. Though these issues are scientific matters, they cannot be resolved by instigating more scientific investigations since it is the principles and institutions, through which the necessary scientific judgement is exercised, which are themselves called into question. Such debates could, in principle, persist indefinitely. They are 'resolved' when a compromise is accepted or when one party gives in; they cannot be resolved simply by finding definitive evidence since there is no such thing. At present, one possible compromise is the introduction of labelling so that European consumers, if they wish, can avoid GM products. But, as discussed in Chapter 4, this labelling initiative may itself be challenged since North American exporters claim that the labels warn consumers off a product that is – they assert – every bit as safe as conventional foods. This issue cannot be resolved by more scientific investigation either.

A second irony is that while these disputes are carried on, farming practices persist largely as 'business as usual'. As was shown in Chapters 5 and 6, the planning system in the UK is designed to consider, on a case-by-case basis, whether some development can be allowed or not. There is no forum for deciding what the best development would be. The question is only: should this development be allowed to go ahead or should it be stopped? Some authors have argued that part of the problem with the GM debate is that it is being considered in this planning context. They have sought to frame the question about GMOs in a different way. Rather than asking are GM crops safe enough to be allowed to be planted, they have asked what would the best form of future for agriculture be? (Stirling and Mayer 1999). Predictably enough, what they find is that – depending on whom they ask – they receive strongly contrasting views on what the best form of agriculture for Britain would entail (Yearley 2001). However, what they do find agreement about is that current agricultural practices are undesirable. Too many agrochemicals are used; the system depends on unsustainable subsidies and petrochemical inputs. We have mechanisms only for approving or rejecting new technologies on a piecemeal basis; there are no mechanisms for taking a holistic view of the best overall strategy. A more comprehensive form of technology assessment would demand that questions are asked of today's practices also, and this might hasten compromise on the otherwise interminable issue of proving the safety or dangerousness of GMOs.

11
The Value of Environmental Sociology: Towards a Sociology of the Sustainable Society

In place of a conclusion

This book is primarily a collection of empirical studies in environmental sociology. Each of the studies has its own analytical conclusions and it is not the role of this chapter either to list those conclusions again or to offer general observations that would direct attention away from the specific focus of each study. Rather, my aim in this chapter is to do something a little different. For over a decade now, as noted in Chapter 10, policy analysts, NGOs and government officials have spoken of sustainable development as a standard or as an ideal for environmental policy and action in the future. In the famous words of the Brundtland Commission, the United Nations committee that first brought the phrase to international prominence, sustainable development is a form of development that meets the needs of the present without compromising the ability of future generations to meet their own needs (World Commission on Environment and Development 1987, 43). The discussion in this concluding chapter will focus on the distinctive contribution that sociology can make to the conceptualisation of sustainable living.

The first observation prompted by sociological studies of environmental issues is that visions of sustainability to date have been unjustifiably restricted. They have generally been concerned with technical matters, with economic arrangements or with questions of political values and ideology; but the attention paid to social institutions and social practices has been very slight. Key questions about the operation and constitution of sustainable societies have simply not been asked. Accordingly, in this final chapter, I set out to ask not 'what should the ethical values of a sustainable society be?' nor 'how would people in a sustainable society generate their energy?', rather 'how would an intendedly sustainable society operate?'. In other words, I propose to shift attention to the classical sociological 'problem of order' – how is society to be enacted and reproduced – in order to examine how the problem of order might be managed in an environmentally sustainable society.

The principal issues at stake can be clarified using an analogy. The architects of communism (or state socialism) offered attractive arguments about the industrial and the economic basis of the new social order. People's needs could be assessed and fairly met; the wastefulness of competitive capitalism could be avoided. But advocates of political change were far less good at anticipating how the society would function: what the culture of industrial workers would be, how the people charged with planning and regulation would conduct themselves, how workers on collective farms would behave or how ruling elites could manipulate the system to perpetuate their advantages. In many respects, it was these unanticipated features of the society that accounted for its downfall. And if one asks what sorts of factors these are, the answer is clearly that they are sociological ones. The aim of this chapter is to use the insights generated from empirical studies of the cultures of environmentalism to ask similar questions for the case of environmentally benign future. Such an enquiry is admittedly exploratory but it is innovative, making a distinctive contribution both to sociology and to social-scientific understandings of environmental issues (see Garforth [2002] for a comparable study of this sort conducted through the analysis of utopian ecological writings).

Sustainability without sociology

To date, the primary emphasis in analyses of sustainable development has been on the technical, the economic and the political-values aspects of sustainability. In many ways this is understandable: one can readily argue that the technical 'baseline data' and infrastructure are indispensable to any approach to sustainable living. One needs to know what the stocks of natural resources are before one can plan how they are to be distributed. Thus, one cannot envisage a sustainable future with specific entitlements to clean water unless one knows how much rainwater there is and where it falls, what artesian reserves there are, how feasible various desalination options may be and so on. From at least the time of the Limits to Growth 'scenarios' work in the early 1970s – work that sought to identify the likely ecological limitations on current ways of producing and consuming – inquiry has focused on how much the earth's natural resources are being run down, what alternatives are conceivably available, and how the transition to any future state could be accomplished. For example, in the field of energy resources, analytic effort has gone into working out how much energy could be saved by conservation and energy efficiency, how much could be produced by benign sources, and how quickly research on new sources would have to mature in order to introduce substitutes before stocks run out or we impose too great a burden on the world through pollution from fossil fuels.

Though the significance of such work is not in doubt, it has two important shortcomings. First, these technical enquiries have commonly been made in isolation from consideration of the mechanisms by which innovations would be implemented and by which resources would be distributed. Even if we knew how much electrical energy was available from photovoltaic cells (cells that produce electricity from sunlight) we still would not know much about how that energy would be owned, traded or used. Second, there is a problem because such exercises, in which technological futures are envisaged and in which future resource-levels are calculated, commonly minimise the significance of technical uncertainties and draw attention away from important issues concerning the 'politics' of knowledge. As the dispute over genetically modified organisms (GMOs), discussed in Chapters 4 and 10 indicates, estimations of the impact of a technology (in this case, future agricultural productivity levels) depend on much more than purely technological factors; projections for the future will differ radically depending on which experts are consulted. Equally, there are persistent difficulties in estimating the stocks of natural resources. It often seems to cynical observers that 'proven' oil reserves grow roughly as quickly as resources are consumed, so that warnings about resource depletion seem always to lose their bite (Bowden 1985; Dennis 1985). In projections about technological futures technical and sociological factors are in practice inseparable. Furthermore, attempts to model the impact of future rainfall patterns or of the spread of photovoltaic cells may be performed in a very 'top-down' way and be insensitive to the intricacies of particular localities or regions, as we saw was the case with aspects of the air-quality modelling described in Chapter 7. Models supposedly concerned with the future of technological systems typically end up trading on implicit assumptions about how people will behave, and those assumptions are anchored in the way people behave today, not how they will behave in the future.

More recently, environmental economists have been among the leaders in defining the concept of sustainable development; indeed, as David Pearce breezily notes, 'sustainable development is readily interpretable as non-declining human welfare over time' (1991, 1). Elsewhere Pearce *et al.* (1989, 48) observe that economists tend to hold one or other of two views about sustainable development. They regard it as equivalent to either passing on to the next generation at least the same capital as the present generation inherited, or passing on the same 'natural capital'. The latter view implies that 'the stock of environmental assets as a whole should not decrease' while the former allows humanly made capital to be traded for natural capital (improvements in health-care provision or increases in the stock of fine buildings might be exchanged for declining areas of forest for example). While there are significant differences between these interpretations, the crucial point is that both depend on attaching economic values to environmental goods; otherwise there is no way of telling whether the capital stock (natural or natural-plus-human) is declining or not. However, for environmental economics to be the neutral medium for assessing sustainability, it is necessary that precise economic values be attached to the environment, and it is in this context that some of the most acute problems occur (see Jacobs 1994). For example, if various measures of value produce discrepant figures, then there are no agreed environmental prices and non-declining capital cannot be determined impersonally.

Worse, the economist's definition of sustainability as non-declining capital does not address distributional issues in the present, and thus does little to resolve transnational disputes about 'rights' of access to environmental resources as noted in Chapters 3 and 4. Finally, though environmental economists appear to have an internally consistent micro-economics of the sustainable economy, it does not follow from this that the system they envisage can function sustainably at a macro-economic level. As the UK governmental economic adviser Tom Burke has pointed out: 'Macroeconomic questions, such as the right level of inflation or the right price of money or the right volume of trade or the right rate of growth, for sustainable development, are vital questions to environmental policy making', while 'the microeconomics of market instruments is pretty marginal to influencing environmental outcomes' (1995, 15). Environmental micro-economics does not guarantee that a sustainable market economy is viable in the long term. Nor is this economistic focus restricted to mainstream approaches. Even alternative measures of sustainability, such as 'ecological footprints' (which attempt to measure how much of the earth's product – expressed in land surface equivalents for growing food, for tree-planting to absorb carbon from the atmosphere and so on – is consumed by different lifestyles) or earlier notions of 'environmental space' remain highly technocratic. To carry out these assessments, people are allocated shares in the earth's product but the studies do not deal with the kinds of governance or popular participation that would promote adherence to these agreements; they do not talk about the kinds of notions of worth and desert that would legitimise the distribution of resources within future societies. There is no explicit and well-developed notion of the moral economy which would sustain the anticipated way of life (see Yearley 1996a, 124–30, 1996b).

Finally, many writers on green political theory have sought to outline the values which could underlie future sustainable ways of living – values such as sufficiency, decentralisation and fairness to other people and other biological systems. Not only do these theorists tend to disagree among themselves about the exact identity of the appropriate values but, even where they agree, it is still notable that they have relatively little to say about the practicalities which would underlie these political values. They seldom give 'thick descriptions' of societies in which these values predominate (on this point see Garforth 2002), leading us directly back to the shortcomings made so clear in the analogy with the case of the Soviet Union.

Overall, the situation of technical and economic approaches to sustainability can be understood, I suggest, in relation to what the Brundtland Report refers to as sustainable development in a 'physical sense' (World Commission on Environment and Development 1987, 43). Brundtland introduces this notion as a 'bottom line'; the argument is not just that 'we' ought to live sustainably, but that in the long run we cannot live any other way. It is, so the argument runs, objectively necessary to become sustainable. It is an inescapable, global imperative. In this way the discourse of sustainable development has become influential very quickly. It is a powerful discourse, from which it is hard to distance oneself since it appears that no one could reasonably disagree with the objective of living sustainably. Yet, this sense of physical compulsion does not tell us if it is possible to live in a sustainable way; it does not tell us how and why to live sustainably, and it certainly tells us next to nothing about the operation of social institutions and social values in a sustainable society. The social institutions of sustainable societies are missing – or are implicit and unexamined – in all three dominant approaches.

Aspects of the sociology of sustainable societies

The empirical studies described in this book – along with other empirical studies in environmental sociology – provide the initial basis for setting out the thick description of sustainable living that the other approaches omit. These issues can be helpfully presented under three sub-headings.

Sustainable values

At first sight it may seem strange to argue that values and morals have been neglected in the treatment of sustainable development. Clearly, a very large part of the argument for moving towards sustainability is precisely ethical: we should not store up for our descendants problems that we cannot ourselves solve or whose solutions we cannot realistically foresee. Equally, a great deal of 'green' writing – both academic and that arising from the movement – focuses on the moral and political foundations of ecologism (see for example Dobson 2000; Carter 2001). Such enquiries are very important and it is not part of my argument to contradict them. Rather, to use the analogy employed above, just as attempts to implement a communist society led to the emergence of everyday value orientations at odds with the avowed aims of the philosophy and just as the design of institutionalised communism led to unintended consequences, so too sociology leads one to ask about the practical value implications and unintended consequences of moves to sustainability. To date, most studies have concentrated on the political philosophy of green values; empirical studies can also examine the sociological implications of those values.

A persistent problem in the Soviet Union, for instance, was that workers subverted communal values by 'borrowing' materials from the work place and by putting in the greatest effort on projects of immediate benefit to their families. Plans supposedly directed at the common good were undermined by individuals' 'rational' adaptations to perceived inefficiencies. The studies in this book do not indicate anything quite as directly practical as this. However, there are significant value strains that relate to moves towards sustainable forms of social organisation. For example, participatory democratic procedures are commonly seen as in tune with the values of ecologism. Green parties sometimes organise their constitutions so as to reflect this value orientation. However, as was indicated in Chapters 2 and 8, environmental organisations and environmental campaigns often cannot be democratic. Demands for democracy are counter-balanced by the requirements of expertise, dedication and professionalism. Similarly, when it comes to international campaigning, the North's environmental groups are typically wealthier and more effective campaigners than those from the South so that problems over representativeness repeatedly arise. The universalist and transnational value ambitions of the environmental movement may not be readily realised – in fact they may even be contradicted – in practice.

Public knowledge, innovation and cognitive legitimacy

It seems clear that knowledge about the requirements of sustainable development is going to have to play a central role in a society which is to be run sustainably. Expertise will be more important even than it is today since experts on sustainability will have to make decisions on issues which are currently delegated to the market under liberal capitalism. If technical knowledge of environmental issues were uncontentious and disinterested and if citizens trusted the views of experts then this might be straightforward. Experts would be invited to specify what the sustainability limits are (for example, what air pollution can specific habitats tolerate) and society would operate within a margin of safety. However, in contemporary highly individualistic societies, official pronouncements which conflict with people's perceived interests are routinely questioned and even taken to court, as we saw was the case with industrial chemicals, planning decisions and the case of the testing of GM crops (see also Turner 2001). The inability of official experts to settle public policy matters in an agreed way, even when there are regulations and standards, was clearly shown in the case of the Brent Spar. Worse still, much of the scientific and technical information needed for environmental decision-making is open to varying interpretations as was described in Chapter 8. Estimates of oil reserves, of fish stocks, of biodiversity and of a host of other environmental touchstones are open to conflicting interpretations even within the relevant expert communities. In contentious areas – as argued in Chapter 10 – it will typically not even be possible to agree about the identity of the relevant experts to advise political decision-makers.

The key issue therefore is that sustainable societies will make unprecedented calls on expertise but, at the same time, the very issues over which expertise will need to be most authoritatively exercised are the ones most open to question. Some commentators have argued that 'social experiments' in community-based problem-solving could be part of the solution to this difficulty. Such activities may tend to have fewer problems with legitimacy than officially appointed experts and they may be more flexible in fusing social and technical elements of remedies (Irwin *et al.* 1994). This implies that a key issue in envisaging 'thick' sustainable futures will be the development of ideas for innovative forms of public expertise and novel institutions which enjoy cognitive authority. A practical intervention involving public expertise was described in Chapter 7.

In addition to demands for expert regulatory insights and advice, most envisionings of a sustainable society place considerable reliance on innovation: for new energy sources, for greater efficiency in resource use and so on. Yet the question of how to run a sustainable science and technology policy is rarely addressed in the environmentalist literature even though – as Chapter 10 indicated – innovative foodcrops gave rise to the largest recent public environmental controversy. The way that GMOs were handled perhaps indicates how to run an unsustainable approach to innovations, but this case clearly also poses problems for mainstream approaches to environmental futures that take technological innovation for granted.

Sustainability for what?

As commonly presented, sustainable development refers to 'getting by' – it is a state in which there is no further diminution of environmental resources. Environmental campaigners often present sustainability as their goal; sustainable development appears to them a lofty achievement towards which society must be shepherded. However, it is possible to take almost the opposite position. One might say that sustainability is a minimum threshold which societies must reach. In a strict sense, unless societies become sustainable they will decline – sustainability is thus not a towering aim but the lowest conceivable attainment. While it is probably more politically expedient to present sustainability as the prize rather than a mere entry ticket, the 'contrary' position does helpfully encourage us to ask: what is sustainability for? In other words, while the achievement of sustainable development may at present be a lifetime's goal, once one has secured sustainability, sustainable development will cease to be an end in itself and will open novel questions about how to live well and about the socially sanctioned aims of life.

The third question about 'thick' accounts of sustainability therefore deals with the question of the aims and ends of the sustainable society. Existing market economies depend on 'betterment' and 'distinction' as motivating forces (Bourdieu, 1984). Moreover, in high-modern cultures consumption appears to have moved even further from simple provisioning (providing the necessities of life) to an apparently autonomous activity in its own right. Going shopping is perhaps the leading leisure activity in Britain, Japan and many parts of North America. Consumers seem to favour confirming (perhaps even 'constructing') their individuality through their purchasing decisions and accordingly the market for 'designer' goods and for apparel bearing the maker's label has swollen dramatically. Ecological concerns have not meshed easily with this process to date since environmentalists' favourite arguments about sufficiency appear to miss the mark altogether. The shopping experience (Falk and Campbell, 1997), precisely because it is no longer about getting the most goods for the least money, is not readily susceptible to the imposition of direct environmental performance standards. In a culture that has swapped consumption and acquisitiveness for sustainability and fairness, it is unclear what the markers of social success will be. The moral economy of such societies – how worth or progress or personal success would be expressed – is not addressed in the standard environmental literature. This realisation invites us to ask whether, and in what ways, the demands of sustainability shape the kinds of goals and objectives it is possible to honour in a sustainable society, and what bases for 'distinction' remain available. The studies in this book indicate that the dynamics of environmental controversy do not encourage reflection on this issue. Disputes tend to be narrowly focused and to take the surrounding sociological milieu for granted. Questions about how to live meaningfully within environmental limits, about what the 'good life' would involve, are constantly marginalised.

In all, therefore, I suggest the studies in this volume demonstrate the value of environmental sociology in two contrasting ways: the concrete and the conjectural. Environmental sociology can teach us about how decisions concerning the environment in fact get made today but it can also encourage us to reflect in a novel way on the nature of environmental futures. Sociology alone is, of course, not sufficient for this task. But in this chapter, I hope to have shown that the neglect of sociology has meant that envisionings of environmental futures have too often been unrealistic and nowhere near 'thick' enough in the understanding of the social fabric of sustainable living.

Notes

2 Social Movement Theory and the Character of Environmental Social Movements

1. I am very grateful to the stimulus given to this chapter by the invitation to give one of the Linacre Lectures 2000 at Linacre College, Oxford. That lecture was subsequently written up as a chapter in the volume of collected lectures (Yearley 2003); it examined the implications of the organisational features of the ecological movement for the likely future of environmentalism in the new century.
2. That is not to say that the argument has not been attempted: neo-Marxists used the notion of a 'permanent arms economy' to propose that military spending was functional for capitalism. On this view, the military consumed without producing and thus was a crucial influence on the demand side of the economy, particularly during recessionary times. Though this argument made some sense in the US case, the evidence appeared to indicate that such spending was precisely not functional for the success of the capitalist economy in the longer run.

3 Shell, a Sure Target for Global Environmental Campaigning?

1. *The Ecologist* magazine contains a 'Campaigns & News' feature in each issue; no details of authorship are given. Information from this source has been referenced in the text, not in the references in the full bibliography.
2. Interviewees have been anonymised. Quotes are taken from recorded interviews and 'character' labels 'Int' for interviewer and R1 for first respondent and so on are given only where they are needed for clarity's sake. R1 and R2 do not necessarily refer to the same persons throughout.
3. The Braer and Sea Empress are both oil tankers which caused spills in Britain, in Scotland in 1993 and in Wales in 1996 respectively. The Sea Empress spill affected the area around the Milford Haven oil terminal in Wales.
4. The G7 refers to the meeting of what were the seven largest economies in the world: the USA, Japan, Germany, the UK, France, Italy and Canada. It becomes the G8 when Russia is present. These countries are slowly being overtaken by China and India in terms of their overall annual income (China's population is after all several tens of times the size of Canada's). At present the G7 shows no signs of changing its make-up and may end up becoming sidelined.

5 Bog Standards: Contesting Conservation Value at a Public Inquiry

1. The research on which this paper is based was supported by an ESRC (Economic and Social Research Council) grant (A00069295). I should like to express my gratitude to Brian Wynne for his detailed comments on this paper.
2. The Giant's Causeway is a celebrated geological feature on Northern Ireland's north coast where hexagonal basaltic columns run from the cliff into the sea; it was earlier thought of as a causeway used by superhuman figures to travel between Ireland and Scotland. Strangford Lough is a shallow marine loch (lough) on the east coast, rich in wildlife and magnificent views and largely free of giants.
3. This has long been an issue in Irish natural history; for historical examples see Whyte (2000).
4. To avoid the endless repetition of 'Northern Ireland' I shall occasionally use the term 'province' even if this might be found politically objectionable by some readers. Within the UK, Northern Ireland is – technically speaking – a Province. But to call it that rather implies that Northern Ireland properly belongs to the UK, something which is far from widely agreed. None the less, in Northern Ireland one can generally talk about 'the province' without causing too much offence.
5. This is a useful point to note some details of my method in this investigation. I attended the inquiry throughout, collected all the supporting written material I could, interviewed a number of the conservationist participants and made three field visits, in company with various environmentalists, to the peat bog itself. I also checked the details of my analysis with my respondents to make sure that I had understood various technical points correctly.
6. The previous designation, the Area of Scientific Interest (ASI), was a limited measure, essentially offering grounds for the DoENI to oppose planning applications if they saw fit. The ASI designation was commonly held to be too weak, see Newbould (1987, 82).
7. See Article 24 of The Nature Conservation and Amenity Lands (Northern Ireland) Order 1985 (London: HMSO, 1985, 18–19).
8. See paragraph 7 of the CWB's submission to the Inquiry.
9. See ibid., appendix 3a.
10. See ibid., paragraph 10.3.
11. In his semi-autobiographical account of scientific conservation work, N. W. Moore recounts that when trying to select the best Mendip ash-wood to adopt as a National Nature Reserve 'It was tempting to develop a scoring system, but since there was no common denominator such a system could do no more than quantify our opinions' (Moore 1987, 79). It seems that when the selection criteria are not going to be challenged it is safe to depend on one's naked judgement. However, erecting a scoring system does little to ward off the determined adversary as the developer's QC demonstrated.
12. It was also important to the legal argument to establish that the need to declare the ASSI had been made clear to the DoENI at the time and I am happy to report that a letter I wrote to the head of the environment service

formed part of this case. It turned out that my days of bog enthusiasm at the inquiry and on Ballynahone More were not passed aimlessly, even if my role was exceptionally minor.

6 Independence and Impartiality in Legal Defences of the Environment

1. The RSPB is Britain's (indeed Europe's) largest organisation devoted solely to nature conservation or environmentalism with – depending on how exactly you count them – around a million members. It works to protect birds and their habitats in Britain and elsewhere; see the centenary publication, Samstag (1988) and also Yearley (1992a).
2. The RSNC was very influential in the foundation of the NCC's forerunner, the Nature Conservancy; see Lowe and Goyder (1983, 154–57).
3. The UK consists of GB and Northern Ireland. For most of the last three decades of the twentieth century the latter was under direct rule from London since its own parliament had become discredited. It retains its own legal system, civil service and other institutions. In GB the NCC stands (or rather, stood since it has been reformed and renamed) outside the civil service; an independent body of this sort was not formed in Northern Ireland where its functions were performed at the time of this case study and of the study in the preceding chapter by a constituent part of the DoENI: the CWB. Campaigners have often argued that this meant that the DoE in Northern Ireland was both 'gamekeeper and poacher'.
4. The ruling rejoices in the name of the European 'Directive on the assessment of the effects of certain public and private projects on the environment' (85/337/EEC).
5. On terminology for naming Northern Ireland, see the end notes to the previous chapter (note 4).
6. As its name implies, the AONB is a designation for areas of exceptional landscape value; its provisions constrain development in those areas.
7. On the British public inquiry system see the last chapter and Wynne (1982, 52–73). As Wynne makes clear (pp. 97–104), a persistent problem for objectors to the nuclear plant that formed the core of his study was their lack of financial resources; by contrast the nuclear industry was supported by government. In the present case also, objectors received no funding to mount their arguments. It should be noted however that certain environmental objectors, such as the NCC and the Council for Nature Conservation and the Countryside (the statutory advisory body in Northern Ireland), would be able to use public funds to hire legal representatives for public inquiries. One might say that questions about impartiality could be raised in such cases because, after all, the government is paying for the presentation of the arguments. However, as we shall see, the issue of impartiality arises much more strongly in the present example where an objector has a commercial relationship with one of the would-be developers.
8. Respectively, these documents were UWT, 'Preliminary Environmental Impact Assessment' (1989a) and 'Proof of evidence to the Killyleagh Marina Inquiry' (1989b).
9. This is minuted in UWT Conservation Committee minutes (CC/60/M, page 3).

10. One anonymous reviewer kindly pointed out that it is not self-evidently necessary for environmental advisers to keep facts and values separate. For instance, on a 'deep green' view, the artificial separation of facts and values is just one of the symptoms of our ecological and spiritual malaise. Conceivably, people could come to deep green experts for factual-cum-evaluation advice. This seems to me to open a vast and fascinating debate. At this point I would like to make only two comments. First, the leading conservation groups in the UK are far from deep green and therefore the problem I describe is also the problem as they perceive it. Second, however justified deep green views are, those views would not currently cut much ice in a British public inquiry. For, as Wynne notes (1982, 130), 'Judicial rationality also assumes the complete separability of facts from values or emotions'. See also Cramer and van den Daele (1987, 17–64).
11. Though the focus for this chapter is on the ideal and practicalities of impartial advice and the fate of such advice in adversarial settings, readers will be interested to know (I hope) that the DoENI turned down all three proposals, although a compromise proposal was subsequently resurrected. By 2003 the Marina plans were still at the proposal stage with conservation groups opposing details of the revised design.
12. As both Harry Collins and Steven Shapin pointed out to me, there is potentially a more general issue here about the social distribution of trust and about the circumstances under which integrity may publicly be called into question or – on the other hand – routinely accepted. For example, many firms of solicitors as well as those 'city' companies employed in the regulation of financial markets are simply assumed to be able to separate out their various functions despite the apparently falsificatory instances of insider dealing. Comparative studies of the role of trust in different professions would be of clear value.

7 Modelling the Environment: Participation, Trust and Legitimacy in Urban Air-Quality Models

1. My work on this topic was supported by two Economic and Social Research Council (ESRC) awards (grant numbers R000221902 and L485274033). I would like to express my thanks to John Forrester and Peter Bailey who worked with me on the first of these grants. I am grateful to them for their permission to use our transcripts in this single-authored work which ties together the concerns of our joint study and my later ESRC award. I would also like to thank Lynn Kilgallon for her invaluable help with the transcribing.
2. Of course, this is not to overlook the importance of trust within the 'lifeworld' of research science, as extensively documented by Shapin (1994); the connection between the two aspects of trust is drawn out in Yearley (1994, 247–55). This point has also been remarked in the field of risk communication, where analysts have noted that the public has a strong interest in weeding out partial and tendentious claims and that knowledge and trust are closely related. But once it is acknowledged that trust in the source of information about a risk or hazard is important to understanding people's response, the typical move among policy makers and students of policy analysis has been to try to break the phenomenon of 'trust' down into its various components.

Psychologists may attempt to model the procedures people appear to use in determining the trustworthiness of various sources, concerning for example the perceived expertise and public interest orientation of the body. Or they may look at the credibility which the public attaches to different types of organisation or different communication media (they may, for instance, ask whether television news is regarded as more credible than official information leaflets). The guiding assumption is that the resulting information can then be used to make scientific organisations more credible and to diminish the likelihood of (irrational) non-acceptance of their pronouncements. In an exactly analogous way, where arguments about trust and credibility have been taken on board in relation to public understanding of science (PUS), the typical response in the scientific community has been to supplement an interest in PUS with a study of factors affecting public trust in science. To put it another way, confident in the correctness of their scientific views, science communicators see public distrust as a distortion, a problem to be overcome; they aim to find approaches which prevent science's signal being disrupted by the noise of distrust. The flaw in this reasoning is its working assumption that trust and credibility are fixed dispositions, either of individuals or of institutions. But trust and credibility are the outcome of interactions and negotiations, not static 'factors' (this argument is developed in Yearley 1999b).

3. I say 'up to seven' because not all aspects of the study were run in all seven cities: Athens, Barcelona, Frankfurt, Manchester, Stockholm, Venice and Zurich; see Shackley *et al.* 1999; Kasemir *et al.* 2003.
4. Subsequently all urban local authorities in England and Wales were required to buy in or develop air-quality modelling capacity and to use these models to anticipate where urban air pollution would still be likely to exceed national thresholds by the end of 2005. The authorities then had to develop Action Plans to try to prevent these pollution 'exceedances' from occurring (see Yearley *et al.* 2003).
5. In the transcription citations, the first figure in square brackets refers to the group number (1 is the city's 'Environment Forum', 4 is the public-sector group and so on; see the methods section earlier in the paper for fuller details of the groups' composition) and the second to the page of the transcript. For anonymity's sake, group members are referred to only by number; the numbering is however consistent, so that – for example – utterances by 'R2' in Group 4 are always by the same individual. Square brackets indicate words switched to preserve a measure of anonymity or other editorial substitutions. Curly brackets indicate actions or other non-verbal occurrences.
6. The identities in this section of transcript are unclear as the quality of recording meant that voices were not fully distinct.

8 Green Ambivalence about Science

1. The transport of the nuclear material to Sellafield was often justified in terms of the 'reprocessing' it would eventually permit. In other words, the idea was that the old fuel would be chemically treated to recover usable atomic material. Spent fuel from other countries (notably Germany and Japan) was also to be imported, reprocessed and then exported again. But problems with the

economics and management of this process which have become apparent in the last decade have dictated that the reprocessing business be scaled down and then indefinitely suspended. The highly expensive reprocessing plant (known as THORP) has barely been used and is scheduled to close in 2010.
2. A photograph of the collision can be seen in Allaby's book (1989, 127); significantly for my present argument, the book is entitled *Green Facts.*
3. The evidence considered in the following pages comes from two sources: from my observation and interview study of non-governmental environmental groups in Northern Ireland and GB from the late 1980s through to the mid-1990s (initially funded by ESRC award A0925 00061) and from an analysis of the published literature (both by and about NGOs) on issues tackled by leading environmental groups.
4. TBT was statutorily withdrawn from use shortly after this.
5. Of course, these problems are not peculiar to greens. Even chemical companies face similar trade-offs between how much to spend on research as against new plant, marketing, training days and so on. The contrast I am drawing is not between environmentalists and everyone else who uses science, but between environmentalists' hopes of science and the practicalities they experience.
6. This is to simplify an already simplified account. The pattern of flow is rather more complicated than I suggest here, see Fred Pearce (1989, 139–43). In particular, some of the deep-water current rises in the Antarctic.
7. It ran in many papers but see for example *The Independent*, 5 July 1989, 9.
8. There was no author given for this report in *Earth Matters*, Autumn/Winter 1990, 4.

9 Mad about the Buoy: Trust and Method in the Brent Spar Controversy

1. This chapter is based on a talk I prepared for the annual research methods training week-end run by the Sociology Department of the University of Surrey in Bournemouth, 27 November 1999; the theme of the weekend was 'data'. A title I toyed with was (with apologies to Shapin and Schaffer) 'Leviathan and the oil pump'. A contrasting sociological study of the episode is presented in Claire Stevens' PhD, University College, London, 2002, entitled 'Alternative truths?: a study of the use of science by the environmental groups in the UK'; see also Huxham and Sumner (1999).
2. The Greenpeace website dedicated to the Brent Spar case gives a date of 16 February 1995, see http://archive.greenpeace.org/comms/brent/index.html. Rose is circumspect about the date in his book-length account.
3. Quoted from Melchett's letter as reproduced at http://archive.greenpeace.org/comms/brent/sep04.html.
4. This pattern is the one described in Chapter 8, where critics attacked Greenpeace and other campaigning environmental groups for playing 'fast and loose' with scientific information – accepting scientific data when it suited them and dismissing data that failed to suit.
5. At the time of the Brent Spar controversy, digital TV news media were relatively unimportant in the UK; the TV news coverage was dominated by four terrestrial channels, two belonging to the BBC, one to a network of private, regional

suppliers (ITV) that shares a main national news provider (ITN), and one to a nearly nationwide company (Channel 4) whose news coverage tends to aim to compete with BBC2, the BBC's more demanding, less popular channel.

10 Genetically Modified Organisms and the Unbearable Irresolution of Testing

1. This chapter is based on a talk I presented at Williams College, Massachusetts, in November 2000. The research on which it was based was supported by the European Commission, Environment and Climate Change project ENV4970695. I should like to record my thanks to Peter Bailey for his collaboration during this project and to our other European partners also.
2. *Bacillus thuringiensis* occurs naturally on plants and in the soil. Under the right circumstances, if ingested by certain caterpillars, the protein Bt produces will attack their gut lining and eventually kill the caterpillar. Varieties of the microbe have been refined and used in spray form since the 1960s as a natural insecticide. Bt has to be sprayed quite liberally in order to make sure that the caterpillars consume a large-enough dose. When plants are engineered to manufacture it themselves, the caterpillars get the protein dose as they eat the plants.
3. Lectins are a very diverse set of proteins that selectively bond with carbohydrates in cell structures. They are common in plants, particularly in seeds and tubers. People are very intolerant to some lectins, for example those in red kidney beans. Lectins may cause sickness by reacting with the gut wall. Other lectins may even pass through the gut wall and disruptively affect other body organs.
4. It was subsequently argued that the simulations in which this effect was produced were unrealistic since it was extraordinarily likely that enough material would be carried on to the insects' preferred food plants to have any effect on the butterfly population. More butterflies, it was said, were likely to be harmed by car windscreens than by this circuitous route. Pimentel and Raven (2000) assess other threats to the butterflies, notably from habitat loss and from other agrochemical treatments.
5. See http://www.defra.gov.uk/environment/acre/advice/advice01.htm for ACRE's advice to the Secretary of State, 23 June 1998.
6. Formerly at www.royalsoc.ac.uk/press/pr_15_99.htm; now see http://www.royalsoc.ac.uk/gmplants/.
7. On the related matter of how the 'ethical dimensions' of GM have become polarised see Wynne (2001).
8. In England such signs typically announce that 'trespassers will be prosecuted'. This is widely enough known that in the Winnie-the-Pooh stories there is a broken sign saying 'Trespassers W' whose meaning is clear to all readers but not to Pooh and friends who believe that the sign commemorates Piglet's grandfather, presumed to have been called Trespassers William.

Bibliographical References

Agarwal, Anil and Sunita Narain (1991) *Global Warming in an Unequal World: A Case of Environmental Colonialism*, Delhi: Centre for Science and Environment.

Allaby, Michael (1989) *Green Facts*, London: Hamlyn.

Allen, Robert (1992) *Waste Not, Want Not*, London: Earthscan.

Allen, Robert and Tara Jones (1990) *Guests of the Nation: The People of Ireland Versus the Multinationals*, London: Earthscan.

AURIS (1994) *Removal and Disposal of the Brent Spar: A Safety and Environmental Assessment of the Options*, Aberdeen: AURIS.

Barnes, Barry (1985) *About Science*, Oxford: Blackwell.

Beck, Ulrich (1992) *Risk Society: Towards a New Modernity*, London: Sage.

Benedick, Richard E. (1991) *Ozone Diplomacy: New Directions in Safeguarding the Planet*, Cambridge, MA: Harvard University Press.

Bennie, Lynn G. (1998) 'Greenpeace and the Oil Companies: Beyond Brent Spar', *Parliamentary Affairs*, 51: 397–410.

Berger, Peter L. (1987) *The Capitalist Revolution: Fifty Propositions about Prosperity, Equality, and Liberty*, Aldershot, Hants: Wildwood House.

Böhme, Gernot, Wolfgang van den Daele and Wolfgang Krohn (1976) 'Finalization in science', *Social Science Information*, 15: 307–330.

Bourdieu, Pierre (1984) *Distinction: A Social Critique of the Judgement of Taste*, London: Routledge & Kegan Paul.

Bowden, Gary (1985) 'The social construction of validity in estimates of US crude oil reserves', *Social Studies of Science*, 15: 207–40.

Bramble, Barbara J. and Gareth Porter (1992) 'Non-governmental organizations and the making of US international environmental policy', in Andrew Hurrell and Benedict Kingsbury (eds) *The International Politics of the Environment*, Oxford: Clarendon Press, 313–53.

Bullard, Robert D. (1990) *Dumping in Dixie: Race, Class, and Environmental Quality*, Boulder, CO: Westview.

Burke, Tom (1995) 'View from the inside: UK environmental; policy seen from a practitioner's perspective', in Tim S. Gray (ed.) *UK Environmental Policy in the 1990s*, Houndmills & London: Macmillan, 11–17.

Carter, Neil (2001) *The Politics of the Environment: Ideas, Activism, Policy*, Cambridge: Cambridge University Press.

Collingridge, David and Colin Reeve (1986) *Science Speaks to Power: The Role of Experts in Policymaking*, New York: St Martin's Press.

Collins, Harry M. (1988) 'Public experiments and displays of virtuosity: the core-set revisited', *Social Studies of Science* 18: 725–48.

Collins, Harry M. (1992) *Changing Order: Replication and Induction in Scientific Practice*, Chicago: University of Chicago Press.

Corporate Watch (1997) No. 4, June.

Cramer, Jacqueline (1987) *Mission-Orientation in Ecology: The Case of Dutch Fresh-Water Ecology*, Amsterdam: Rodopi.

Cramer, Jacqueline and Wolfgang van den Daele (1987) 'Is ecology an "alternative" natural science?' in Jacqueline Cramer (ed.) *Mission-Orientation in Ecology: The Case of Dutch Fresh-Water Ecology*, Amsterdam: Rodopi, 17–64.

Cruickshank, Margaret (1987) 'Peatlands of Ireland', in Frank Mitchell (ed.) *The Book of the Irish Countryside*, Belfast: Blackstaff Press, 106–12.

della Porta, Donatella and Mario Diani (1999) *Social Movements: An Introduction*, Oxford: Blackwell.

Dennis, Michael A. (1985) 'Drilling for dollars: the making of US petroleum reserve estimates, 1921–25', *Social Studies of Science*, 15: 241–65.

Department of the Environment (NI) (1984) *North Down and Ards Area Plan, 1984–1995*, Belfast: HMSO.

Dickson, Lisa and Alistair McCulloch (1996) 'Shell, the Brent Spar and Greenpeace: a doomed tryst?' *Environmental Politics*, 5: 122–29.

Dobson, Andrew (2000) *Green Political Thought*, London: Routledge.

Dowie, Mark (1995) *Losing Ground: American Environmentalism at the Close of the Twentieth Century*, London: MIT Press.

Dratwa, Jim (2002) 'Taking risks with the precautionary principle: food (and the environment) for thought at the European Commission', *Journal of Environmental Policy and Planning*, 4: 197–213.

Dryzek, John S., David Downes, Christian Hunold and David Schlosberg with Hans-Kristian Hernes (2003) *Green States and Social Movements: Environmentalism in the United States, United Kingdom, Germany and Norway*, Oxford: Oxford University Press.

Earth First! Action Update (1997) No. 41, July and August.

Elleker, Andrew D. (1995) *The Air Quality of Sheffield April 1994 – March 1995*, Sheffield: Sheffield City Council.

Entwistle, Vikki, Mary Renfrew, Steven Yearley, John Forrester and Tara Lamont (1998) 'Lay perspectives: advantages for health research?' *British Medical Journal*, 316 (7 February): 463–66.

Epstein, Steven (1995) 'The construction of lay expertise: AIDS activism and the forging of credibility in the reform of clinical trails', *Science, Technology and Human Values*, 15: 495–504.

Eyerman, Ron and Andrew Jamison (1991) *Social Movements: A Cognitive Approach*, Cambridge: Polity Press.

Falk, Pasi and Colin B. Campbell (eds) (1997) *The Shopping Experience*, London: Sage.

Finger, Matthias and James Kilcoyne (1997) 'Why transnational corporations are organizing to "save the global environment"', *The Ecologist*, 27: 138–42.

Forrester, John (1999) 'The logistics of public participation in environmental assessment', *International Journal of Environment and Pollution*, 11: 316–30.

Friends of the Earth (1992) *Twenty-One Years of Friends of the Earth*, London: Friends of the Earth.

Funtowicz, Silvio O. and Jerome R. Ravetz (1991) 'A new scientific methodology for global environmental issues', in Robert Costanza (ed.) *Ecological Economics*, New York: Columbia University Press, 137–52.

Garforth, Lisa (2002) *Green Utopias: Imagining the Sustainable Society*, DPhil Thesis, York: University of York.

Giddens, Anthony (1990) *The Consequences of Modernity*, Cambridge: Polity Press.

Habermas, Jürgen (1971) *Toward a Rational Society: Student Protest, Science and Politics*, London: Heinemann.
Hannigan, John (1995) *Environmental Sociology: A Social Constructionist Perspective*, London: Routledge.
Hansen, Anders (2000) 'Claimsmaking and framing in British newspaper coverage of the Brent Spar controversy', in Stuart Allan, Barbara Adam and Cynthia Carter (eds) *Environmental Risks and the Media*, London: Routledge, 55–72.
Horsfall, John (1990) 'The hijack of reason', *The Guardian*, 20 April, 25.
Huxham, Mark and David Sumner (1999) 'Emotion, science and rationality: the case of Brent Spar', *Environmental Values*, 8: 349–68.
Irwin, Alan (1989) 'Deciding about risk', in Jennifer Brown (ed.) *Environmental Threats: Perception, Analysis and Management*, London: Belhaven, 19–33.
Irwin, Alan (1995) *Citizen Science: A Study of People, Expertise and Sustainable Development*, London: Routledge.
Irwin, Alan and Brian Wynne (1996) 'Introduction', in Alan Irwin and Brian Wynne (eds) *Misunderstanding Science? The Public Reconstruction of Science and Technology*, Cambridge: Cambridge University Press, 1–17.
Irwin, Alan, Susse Georg and Philip Vergragt (1994) 'The social management of environmental change', *Futures*, 26: 323–34.
Irwin, Alan, Alison Dale and Denis Smith (1996) 'Science and Hell's kitchen: the local understanding of hazard issues', in Alan Irwin and Brian Wynne (eds) *Misunderstanding Science? The Public Reconstruction of Science and Technology*, Cambridge: Cambridge University Press, 47–64.
Jacobs, Michael (1994) 'The limits to neoclassicism: towards an institutional environmental economics', in Michael Redclift and Ted Benton (eds) *Social Theory and Global Environmental Change*, London: Routledge, 67–91.
Jamison, Andrew, Ron Eyerman and Jacqueline Cramer (1990) *The Making of the New Environmental Consciousness: A Comparative Study of Environmental Movements in Sweden, Denmark and the Netherlands*, Edinburgh: Edinburgh University Press.
Jasanoff, Sheila S. (1990) *The Fifth Branch: Science Advisers as Policymakers*, Cambridge, MA: Harvard University Press.
Jasanoff, Sheila S. (1995) *Science at the Bar: Law, Science, and Technology in America*, Cambridge, MA: Harvard University Press.
Jasanoff, Sheila S. (1996) 'Science and norms in global environmental regimes', in Fen O. Hampson and Judith Reppy (eds) *Earthly Goods: Environmental Change and Social Justice*, Ithaca, NY: Cornell University Press, 173–97.
Jasanoff, Sheila S. (1997) 'Civilization and madness: the great BSE scare of 1996', *Public Understanding of Science*, 6: 221–32.
Jasanoff, Sheila S. (1998) 'The eye of everyman: witnessing DNA in the Simpson trial', *Social Studies of Science*, 28: 713–40.
Jasper, James M. (1997) *The Art of Moral Protest: Culture, Biography and Creativity in Social Movements*, Chicago: Chicago University Press.
Kasemir, Bernd, Jill Jäger, Carlo C. Jaeger and Matthew T. Gardner (eds) (2003) *Public Participation in Sustainability Science: A Handbook*, Cambridge: Cambridge University Press.
Keck, Margaret E. and Kathryn Sikkink (1998) *Activists Beyond Borders: Advocacy Networks in International Politics*, Ithaca, NY: Cornell University Press.

Kitsuse, John I. and Malcolm Spector (1981) 'The labeling of social problems', in Earl Rubington and Martin S. Weinberg (eds) *The Study of Social Problems*, New York: Oxford University Press, 198–206.

Krebs, John and John G. Shepherd (1987) 'Disposal of the Brent Spar', *Science and Public Affairs* (no volume) 34–38.

Laird, Frank N. (1993) 'Participatory analysis, democracy, and technological decision making', *Science, Technology and Human Values*, 18: 341–61.

Lambert, Helen and Hilary Rose (1996) 'Disembodied knowledge? Making sense of medical science', in Alan Irwin and Brian Wynne (eds) *Misunderstanding Science? The Public Reconstruction of Science and Technology*, Cambridge: Cambridge University Press, 65–83.

Levidow, Les (1999) 'Britain's biotechnology controversy: elusive science, contested expertise', *New Genetics and Society*, 18: 47–64.

Levidow, Les (2001) 'Precautionary uncertainty: regulating GM crops in Europe', *Social Studies of Science*, 31: 845–78.

Lowe, Philip and Jane Goyder (1983) *Environmental Groups in Politics*, London: Allen & Unwin.

Lowe, Philip and Wolfgang Rüdig (1986) 'Review article: political ecology and the social sciences; the state of the art', *British Journal of Political Science*, 16: 513–50.

Lynch, Michael (2002) 'Protocols, practices, and the reproduction of technique in molecular biology', *British Journal of Sociology*, 53: 203–20.

McCormick, John (1991) *British Politics and the Environment*, London: Earthscan.

McCright, Aaron M. and Riley E. Dunlap (2000) 'Challenging global warming as a social problem: an analysis of the conservative movement's counter-claims', *Social Problems*, 47: 499–522.

McCright, Aaron M. and Riley E. Dunlap (2003) 'Defeating Kyoto: the conservative movement's impact on US climate change policy', *Social Problems*, 50: 348–73.

McKay, George (1996) *Senseless Acts of Beauty: Cultures of Resistance since the Sixties*, London: Verso.

Melucci, Alberto (1989) *Nomads of the Present: Social Movements and Individual Needs in Contemporary Society*, London: Hutchinson.

Merton, Robert, Marjorie Fiske and Patricia Kendall (1956) *The Focused Interview: A Manual of Problems and Procedures*, New York: Free Press.

Millstone, Erik, Eric Brunner and Sue Mayer (1999) 'Beyond "substantial equivalence"', *Nature*, 401: 525–26.

Mol, Arthur (1997) 'Ecological modernization: industrial transformations and environmental reform', in Michael Redclift and Graham Woodgate (eds) *The International Handbook of Environmental Sociology*, Cheltenham, UK & Northampton, MA: Edward Elgar, 138–49.

Moore, Norman W. (1987) *The Bird of Time: The Science and Politics of Nature Conservation*, Cambridge: Cambridge University Press.

Morgan, David (1988) *Focus Groups as Qualitative Research*, London: Sage.

Morgan, David (ed.) (1993) *Successful Focus Groups: Advancing the State of the Art*, London: Sage.

Mulkay, Michael (1979) *Science and the Sociology of Knowledge*, London: Allen & Unwin.

Nelkin, Dorothy (ed.) (1979) *Controversy: Politics of Technical Decisions*, London: Sage.
Newbould, Palmer J. (1987) 'Conservation', in Ronald H. Buchanan and Brian M. Walker (eds) *Province, City and People: Belfast and its Region*, Antrim: Greystown Books, 79–98.
Newby, Howard (1991) 'One world, two cultures: sociology and the environment', *Network* (British Sociological Association newsletter), 50 (May): 1–8.
Nicholson, Max (1987) *The New Environmental Age*, Cambridge: Cambridge University Press.
North, Richard (1987) 'Greenpeace, still credible?' *The Independent*, 21 September, 15.
Oakley, Ann (1972) *Sex, Gender and Society*, London: Temple Smith.
O'Neill, John (1993) *Ecology, Policy and Politics: Human Well-Being and the Natural World*, London: Routledge.
Oteri J.S., M.G. Weinberg and M.S. Pinales (1982) 'Cross-examination of chemists in drug cases', in Barry Barnes and David Edge (eds) *Science in Context: Readings in the Sociology of Science*, Milton Keynes: Open University Press, 250–59.
Pearce, David (1991) 'Introduction', in David Pearce (ed.) *Blueprint 2: Greening the World Economy*, London: Earthscan, 1–10.
Pearce, David, Anil Markandya and Edward B. Barbier (1989) *Blueprint for a Green Economy*, London: Earthscan.
Pearce, Fred (1989) *Turning up the Heat*, London: Paladin.
Pearce, Fred (1991) *Green Warriors: The People and the Politics Behind the Environmental Revolution*, London: Bodley Head.
Pereira, Ângela Guimarães, Clair Gough and Bruna De Marchi (1999) 'Computers, citizens and climate change: the art of communicating technical issues', *International Journal of Environment and Pollution*, 11: 266–89.
Pimentel, David S. and Peter H. Raven (2000) 'Commentary: Bt corn pollen impacts on non-target Lepidoptera; assessment of effects in nature', *Proceedings of the National Academy of Sciences of the United States of America*, 97 (July 18): 8198–199.
Porritt, Jonathon (1989) 'Green shoots, rotten roots', *BBC Wildlife Magazine*, 7: 352–53.
Porritt, Jonathon and David Winner (1988) *The Coming of the Greens*, London: Fontana.
Pye-Smith, Charlie and Chris Rose (1984) *Crisis and Conservation: Conflict in the British Countryside*, Harmondsworth: Penguin.
Rainforest Action Network (1998) 'Action Alert 135'.
Rice, Tony and Paula Owen (1999) *Decommissioning the Brent Spar*, London: Spon Press.
Ringquist, Evan J. (2000) 'Environmental justice: normative concerns and empirical evidence', in Norman J. Vig and Michael E. Kraft (eds) *Environmental Policy*, Washington, DC: CQ Press, 232–56.
Robertson, Roland (1992) *Globalization: Social Theory and Global Culture*, London: Sage.
Robertson, Roland and Frank Lechner (1985) 'Modernization, globalization and the problem of culture in world-systems theory', *Theory, Culture and Society*, 2: 103–17.
Rose, Chris (1993) 'Beyond the struggle for proof: factors changing the environmental movement', *Environmental Values*, 2: 285–98.
Rose, Chris (1998) *The Turning of the Spar*, London: Greenpeace.

Rowell, Andrew (1995) 'Oil, Shell and Nigeria: Ken Saro-Wiwa calls for a boycott', *The Ecologist*, 25: 210–13.

Rowell, Andrew (1997) 'Crude operators: the future of the oil industry', *The Ecologist*, 27: 99–106.

Rüdig, Wolfgang (1986) 'Nuclear power: an international comparison of public protest in the USA, Great Britain, France and West Germany', in Roger Williams and Stephen Mills (eds) *Public Acceptance of New Technologies: An International Review*, London: Croom Helm, 364–417.

Rüdig, Wolfgang, Lynn G. Bennie and Mark N. Franklin (1991) *Green Party Members: A Profile*, Glasgow: Delta Publications.

Sachs, Wolfgang (1994) 'The blue planet: an ambiguous modern icon', *The Ecologist*, 24: 170–75.

Samstag, Tony (1988) *For Love of Birds: The Story of the RSPB*, Sandy, Bedfordshire: RSPB.

Schmidheiny, Stephan (with the Business Council for Sustainable Development) (1992) *Changing Course: A Global Business Perspective on Development and the Environment*, London: MIT Press.

Sclove, Richard E. (1995) *Democracy and Technology*, New York: Guilford Press.

Shackley, Simon (1997) 'Trust in models? the mediating and transformative role of computer models in environmental discourse', in Michael Redclift and Graham Woodgate (eds) *The International Handbook of Environmental Sociology*, Cheltenham: Edward Elgar, 237–60.

Shackley, Simon, Éric Darier and Brian Wynne (1999) 'Towards a "Folk Integrated Assessment" of Climate Change?' *International Journal of Environment and Pollution*, 11: 351–72.

Shapin, Steven (1984) 'Pump and circumstance: Robert Boyle's literary technology', *Social Studies of Science*, 14: 481–520.

Shapin, Steven (1994) *A Social History of Truth: Civility and Science in Seventeenth-Century England*, Chicago: University of Chicago Press.

Shapin, Steven (1996) *The Scientific Revolution*, Chicago: University of Chicago Press.

Shapin, Steven and Simon Schaffer (1985) *Leviathan and the Air-Pump: Hobbes, Boyle, and the Experimental Life*, Princeton, NJ: Princeton University Press.

Sheail, John (1976) *Nature in Trust: The History of Nature Conservation in Britain*, Glasgow: Blackie.

Sheail, John (1987) *Seventy-Five Years in Ecology: The British Ecological Society*, London: Blackwell.

Silvertown, Jonathan (1990) 'Earth as an environment for life', in Jonathan Silvertown and Philip Sarre (eds) *Environment and Society*, London: Hodder & Stoughton, 48–87.

Smith, Roger and Brian Wynne (eds) (1988) *Expert Evidence: Interpreting Science in the Law*, London: Routledge.

Stirling, Andy and Sue Mayer (1999) *Re-Thinking Risk: A Pilot Multi-Criteria Mapping of a Genetically Modified Crop in Agricultural Systems in the UK*, Brighton, Sussex: Science Policy Research Unit.

Sunderlin, William D. (2003) *Ideology, Social Theory and the Environment*, Lanham MD: Rowman and Littlefield.

The Ecologist (1998) 'The Monsanto Files', *The Ecologist*, 28: 249–318.

Thomas, Jeff and Keith Cook (1998) *Brent Spar – A Triumph for Whom?* Open University MSc Science supplement 392141, Milton Keynes: The Open University.

Touraine, Alain (1981) *The Voice and the Eye: Analysis of Social Movements*, Cambridge: Cambridge University Press.
Touraine, Alain (1983) *Anti-Nuclear Protest: Opposition to Nuclear Energy in France*, Cambridge: Cambridge University Press.
Turner, Stephen (2001) 'What is the problem with experts?' *Social Studies of Science*, 31: 123–49.
Ulster Wildlife Trust (1989a) 'Preliminary environmental impact assessment', Belfast: UWT.
Ulster Wildlife Trust (1989b) 'Proof of evidence to the Killyleagh marina inquiry', Belfast: UWT.
United Nations (1993) *The Global Partnership for Environment and Development: A Guide to Agenda 21*, New York: United Nations.
Wallis, Roy (1985) 'Science and pseudo-science', *Social Science Information*, 24: 585–601.
Weber, Max (1964) *The Theory of Social and Economic Organization*, New York: Free Press.
Whyte, Nicholas (2000) *Science, Colonialism and Ireland*, Cork: Cork University Press.
World Commission on Environment and Development (1987) *Our Common Future* (The Brundtland Report), Oxford: Oxford University Press.
World Resources Institute (1990) *World Resources 1990–91*, New York: Oxford University Press.
Wynne, Brian (1982) *Rationality and Ritual: The Windscale Inquiry and Nuclear Decisions in Britain*, Chalfont St Giles: British Society for the History of Science.
Wynne, Brian (1992) 'Misunderstood misunderstanding: social identities and public uptake of science', *Public Understanding of Science*, 1: 281–304.
Wynne, Brian (1994) 'Scientific knowledge and the global environment', in Michael Redclift and Ted Benton (eds) *Social Theory and the Global Environment*, London: Routledge, 169–89.
Wynne, Brian (2001) 'Creating public alienation: expert cultures of risk and ethics on GMOs', *Science as Culture*, 10: 445–81.
Yearley, Steven (1989) 'Bog standards: science and conservation at a public inquiry', *Social Studies of Science*, 19: 421–38.
Yearley, Steven (1992a) *The Green Case: A Sociology of Environmental Arguments, Issues and Politics*, London: Routledge.
Yearley, Steven (1992b) 'Skills, deals and impartiality: the sale of environmental consultancy skills and public perceptions of scientific neutrality', *Social Studies of Science*, 22: 435–53.
Yearley, Steven (1992c) 'Green ambivalence about science: legal-rational authority and the scientific legitimation of a social movement', *British Journal of Sociology*, 43: 511–32.
Yearley, Steven (1993) 'Standing in for nature: the practicalities of environmental organisations', use of science', in Kay Milton (ed.) *Environmentalism: The View from Anthropology*, London: Routledge, 59–72.
Yearley, Steven (1994) 'Understanding science from the perspective of the sociology of scientific knowledge: an overview', *Public Understanding of Science*, 3: 245–58.

Yearley, Steven (1995a) 'The transnational politics of the environment', in James Anderson, Chris Brook and Allan Cochrane (eds) *A Global World? Re-ordering Political Space*, Oxford: Oxford University Press, 209–47.

Yearley, Steven (1995b) 'The environmental challenge to science studies', in Sheila Jasanoff, Gerald E. Markle, James C. Petersen and Trevor Pinch (eds) *Handbook of Science and Technology Studies*, London: Sage, 457–79.

Yearley, Steven (1996a) *Sociology, Environmentalism, Globalization*, London: Sage.

Yearley, Steven (1996b) 'Campaigning and critique: public-interest groups and environmental change', in Fen O. Hampson and Judith Reppy (eds) *Earthly Goods: Environmental Change and Social Justice*, Ithaca, NY: Cornell University Press, 198–220.

Yearley, Steven (1999a) 'Computer models and the public's understanding of science: a case-study analysis', *Social Studies of Science*, 29: 845–66.

Yearley, Steven (1999b) 'What do we mean by "science" in the public understanding of science', in Meinolf Dierkes and Claudia von Gröte (eds) *Between Understanding and Trust: The Public, Science and Technology*, Reading: Harwood Academic, 217–36.

Yearley, Steven (2000) 'Making systematic sense of public discontents with expert knowledge: two analytical approaches and a case study', *Public Understanding of Science*, 9: 105–22.

Yearley, Steven (2001) 'Mapping and interpreting societal responses to genetically modified food and plants – essay review', *Social Studies of Science*, 31: 151–60.

Yearley, Steven (2003) 'Social movements as problematic agents of global environmental change', in Steven Vertovec and Darrell A. Posey (eds) *Globalization, Globalism, Environments, and Environmentalism: Consciousness of Connections; The Linacre Lectures 2000*, Oxford: Oxford University Press, 39–54.

Yearley, Steven (2004) *Making Sense of Science: Understanding the Social Study of Science*, London: Sage.

Yearley, Steven and John Forrester (2000) 'Shell, a sure target for global environmental campaigning?' in Robin Cohen and Shirin Rai (eds) *Global Social Movements*, London: Athlone, 134–45.

Yearley, Steven and Kay Milton (1990) 'Environmentalism and direct rule: the politics and ethos of conservation and environmental groups in Northern Ireland', *Built Environment*, 16: 192–202.

Yearley, Steven, John Forrester and Peter Bailey (2001) 'Participation and expert knowledge: a case study analysis of scientific models and their publics', in Matthijs Hisschemöller, Rob Hoppe, William N. Dunn and Jerry R. Ravetz (eds) *Policy Studies Review Annual 12: Knowledge, Power and Participation in Environmental Policy Analysis*, New Brunswick, NJ and London: Transaction Publishers, 349–67.

Yearley, Steven, Steve Cinderby, John Forrester, Peter Bailey and Paul Rosen (2003) 'Participatory modelling and the local governance of the politics of UK air pollution: a three-city case study', *Environmental Values*, 12: 247–62.

Zald, Mayer N. and John D. McCarthy (1987) 'Social movement industries: competition and conflict among SMOs', in Mayer N. Zald and John D. McCarthy (eds) *Social Movements in an Organizational Society: Collected Essays*, New Brunswick, NJ: Transaction Books, 161–84.

Index

LaVergne, TN USA
29 October 2010
202809LV00001B/39/P